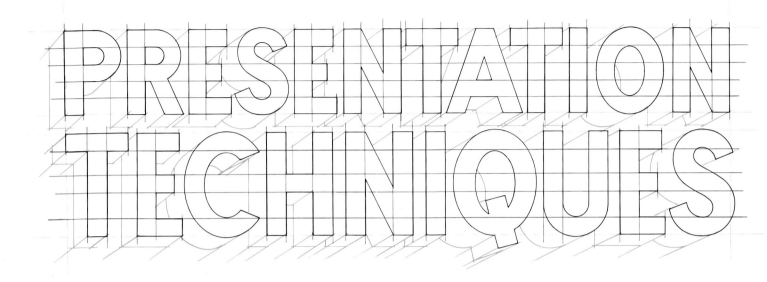

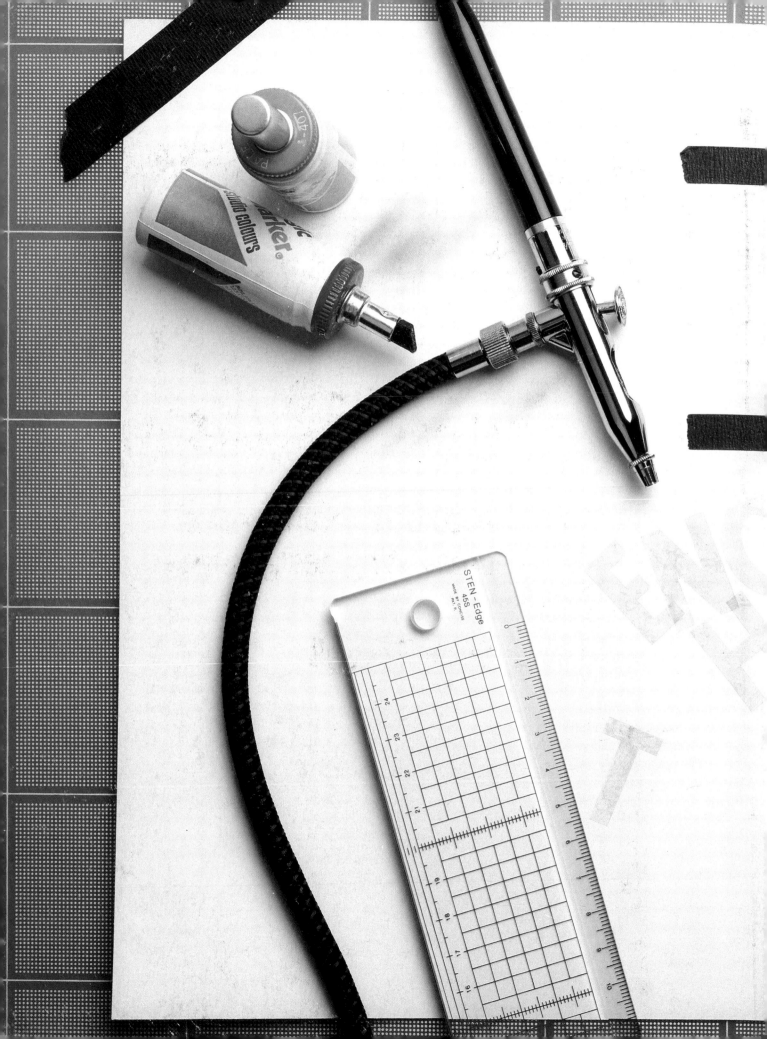

PRESENTATION TECHNIQUES

A guide to drawing and
presenting design ideas

Dick Powell

Orbis · London

For Lucy

Author's Acknowledgments
Special thanks to Ed Holt and Henry Arden for all the photographic sessions, to Richard Seymour for help and support, and to Norbert Linke of General Electric Plastics and Paul Butler and Herry Kleyn van Willigen of Yamaha Motors NV for permission to use their drawings. Special thanks to those people who gave up their time at weekends to create stage-by-stage examples, and to those who spent time sourcing and supplying gallery material.

Thanks also to all the companies who supplied graphics products, especially Royal Sovereign for Magic Markers and 3M for the tapes used in the full-size tape drawings. Finally, thanks to Colin Grant from Orbis and Mike Wade for helping to put the book together.

Picture Credits
All drawings were done by the author except for those credited otherwise, the line drawings on p.14 (coloured by the author), pp.28-39 and pp.49-59, which were done by Geoff Dicks, and the illustrations on pp.152, 153 and 155, which were done by Richard Seymour. The illustration on p.7 was supplied by the Midland Bank plc and that on p.8 top courtesy of the Frank Lloyd Wright Memorial Foundation. Special photography was done by Ed Holt and Henry Arden.

Colour reproduction by Imago Publishing Ltd.

Printed in Italy
ISBN: 0-85613-600-X

Contents

1	Introduction	6
2	Materials	12
3	Perspective Drawing	26
4	Colouring Up	48
5	Marker Rendering	60
6	Airbrush Rendering	86
7	Coloured Paper Rendering	102
8	Automotive Rendering	110
9	Special Finishes	138
10	Descriptive Drawing	140
11	Backgrounds and Mounting	150
12	Conclusion	156
	Index	158

1 Introduction

Design is the creative process which brings ideas to reality. To help meet this challenge, the industrial designer needs to master a wide range of skills from clear, concise report-writing to modelmaking – and all backed up by an extensive understanding of materials and manufacturing processes. But, whatever the level of involvement, from the conceptual thinking behind an innovative new product to the resolution of a difficult moulding problem, the designer will invariably find that, of all these skills, the most important is drawing.

An industrial designer who cannot draw is certainly less efficient and almost always less creative than one who can. Yet, surprisingly, drawing is one of the forgotten subjects of design education, a casualty of the designer's bid for a recognized place in the commercial world. Mistakenly seen by many as an anachronistic indulgence in the fine arts, drawing is *the* basic tool of the industrial designer's trade. Without this skill, too many designers are forced to design only what they can draw, rather than draw what they can design.

Industrial design is a three-dimensional discipline by definition, but unlike a graphic artist who both conceives and executes ideas in two dimensions, the industrial designer must shift first from a three-dimensional idea to a two-dimensional sketch, and then back again to a three-dimensional model. If you can communicate your design ideas well to others, you are also better equipped to communicate them to yourself; you can, so-to-speak, hold a conversation with yourself as you work, shifting quickly and easily from drawing to drawing as an idea develops, and keeping pace with the speed of your own thinking. During the early phases of a design project, this skill allows you to sift through, and sort out, a huge number of potential opportunities, while your less adept colleagues may still be mocking up their first ideas and discovering where they can be improved. Wherever product appearance is an important factor, drawing skills become indispensable in the resolution of complex forms and in bringing them to a stage where they can be modelled. The more a designer exercises these skills, the better he or she will become at product styling, and the better able to visualize and understand ideas as they develop.

It is a hard task indeed to *teach* people to draw; it is, however, an ability that can be developed by careful nurturing and constant practice, particularly by drawing from life. Drawing from life teaches you to look, and to analyse the way you see things, and to understand why the world looks the way it does and, in so doing, understand why one thing can support another, and why some things look more harmonious than others. This book will not be of much use to those who cannot draw at all but is aimed squarely at those who have the ability but lack the basic techniques which can help them to present their ideas to others.

In every design project you tackle, you will need to present your ideas to a client or colleague. For many, presentation can be an intimidating and difficult task, and the less it is done, the more difficult it becomes. What ought to be a natural transition from a working to a presentation drawing becomes a major obstacle to be overcome simply through lack of confidence and ignorance of techniques. This book will bring you up to date with the techniques and, once your confidence is boosted by a little practice, it should be possible to bring about a really significant improvement in the standard of your presentation drawing. One of the best ways of developing such abilities is to analyse finished drawings and see how different techniques are used in practice. For this reason many chapters in this book feature stage-by-stage illustrations, which show in detail how an image is built up, and conclude with gallery sections of finished work, which provide further opportunity for analysis. It should thus prove useful for students and professionals alike, both as basic instruction and as a glossary to refer to when faced with the task of presenting ideas.

History

The 'presentation drawing' goes back to a time when a patron commissioned an artist, craftsman or architect to create an artefact to his specification: a piece of sculpture, a building, a piece of furniture, a monument or whatever. Then, as now, the designer had to show the patron, as best he could, what the end result would look like. Technical drawings apart (which were of course the actual reference for the builders), these presentation drawings started life as pencil sketches and, principally in architecture, developed apace as technical drawing skills progressed using line-work. No real presentation drawings were possible until the advent of perspective during the Renaissance, which, for the first time, gave the patron or client a real chance of understanding an object in three dimensions. It was principally in architecture that these skills were developed with the generation of perspective systems to help the draughtsman accurately define what his building would look like. There have been many books on architectural rendering with some truly superb examples (like the Sir Edward Lutyens/Cyril Farey and Frank Lloyd Wright illustrations shown here), which the serious industrial design student should refer to. It was, however, only with the dawn of the Industrial Revolution that the subject matter of this book began to emerge as a definitive skill.

Wherever goods were to be mass-produced, there were draughtsmen producing the technical drawings from which the tools and dies could be made, and the product produced. The patron, customer, client or manager, however, usually lacked the skill needed to 'read' these drawings and therefore understand what the product would be like *before* it was actually made. Architects

A superb water-colour perspective (1925) by Cyril Farey of Sir Edwin Lutyens's design for the Midland Bank's head office, Poultry, London, completed in 1939.

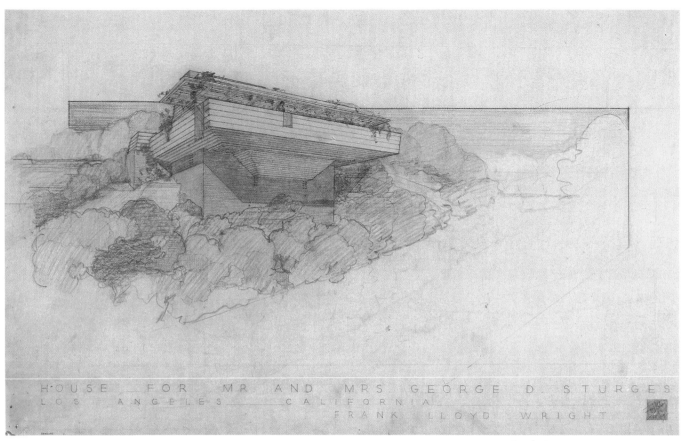

HOUSE   FOR   MR.   AND   MRS.   GEORGE   D.   STURGES
LOS   ANGELES   CALIFORNIA
FRANK   LLOYD   WRIGHT

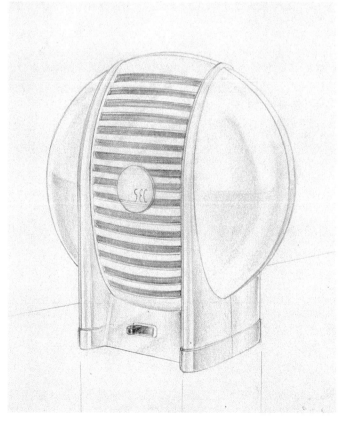

and artists, who had the necessary design and drawing skills, were called in both to improve the appearance of the product and to illustrate it so that others could understand. As the industrial design profession nervously got underway, drawing became central to the designer's role.

We can see the results of the work of these early designers, but few examples of their drawings survive because then, as now, once the product is in the marketplace, the drawing has fulfilled its purpose and can be thrown away. Most early examples were done in pencil or ink, with a water-colour wash, or perhaps, later on, with coloured crayons. Paint was the favoured medium for a long time, and was usually applied with a brush. The airbrush became very popular with designers in the late fifties and early sixties, before the advent of the marker in the mid-sixties. Since then, the marker has dominated, and only looks like being overtaken by the sharp rise in computer graphics. The technical explosion in electronics is making computers accessible to all, and the vast increases in processing power and memory will bring sophisticated drawing programs within the reach of all designers before very long. Once that happens, traditional rendering skills may disappear as the new media take off. However, make no mistake, whether the object in the hand is a coloured pencil, a marker or a computer control, the basic design and visualizing skills needed to use them effectively remain the same.

As well as the techniques and media changing over the years, rendering itself has rather gone out of fashion. The early industrial designers were involved in product design only peripherally, primarily as stylists. They soon discovered, however, that in order to be effective, they needed a deeper understanding of production techniques, engineering, marketing, ergonomics and so on. This experience with, and general appreciation of, all aspects of product development, coupled to a natural creativity, made the designer into an effective problem solver and innovator. As this role developed, designers have found themselves responsible to both client and consumer, sorting out ergonomic, safety and performance factors to satisfy both parties. This new and wider role brought with it a much improved professional status, partly because the ability to rationalize and quantify is more tangible (and, incidentally, more bankable) than the ability to articulate and resolve aesthetic problems. Certainly, this new responsibility gave designers much needed access to the real decision makers,

Above left: *A perspective drawing (1939) by Frank Lloyd Wright of the Sturges House, California, done with pencil and coloured crayons on tracing paper mounted to board. Note how the view breaks out of the rectangular background which also helps to settle the image on the paper.*

Left: *Two presentation drawings by Raymond Loewy, a design for a plug (1939) – far left – and a design for a fan heater (1938) – left. Both pencil on tracing paper mounted to card.*

Above: *A Raymond Loewy design for a moulded fan heater (1938) for GEC. Pencil, gouache and charcoal on grey paper.*

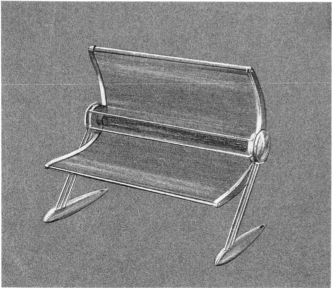

Top: *Sports car for General Motors (1962) – designer unknown. Crayon and gouache on black card.*

Above: *Electric radiant heater by Raymond Loewy (1939). Gouache and coloured crayon on Ingres paper.*

OOBOO

Freightliner

but in the process of acquiring respectability they have tended to lose their early artistic roots. This has been reflected in the colleges where model-making skills have assumed far greater importance than drawing skills.

It is through the model, of course, that an idea reaches three dimensions for the first time, and both designer and client can truly assess the design. A rendering *can never* be a substitute for a model and it is a fool who goes straight to tooling without investing in a model first (many have done it and learnt the hard way!). Good quality models, however, cost a lot of money and take time to produce, making it uneconomical to produce more than one or two. The making of a model can also commit the designer to a layout, proportion and style at a very early stage in the design process; in the eyes of the client, the job appears nearly finished, leaving the designer with limited options. With good quality visuals, on the other hand, the designer can give the client many directions and illustrate specific trade-offs between, for example, cost and appearance. Once a favoured direction emerges the designer can then track as rapidly as possible to models, ensuring at the same time that all the technical, ergonomic and user problems have been correctly addressed.

Partly as a result of the economic downturn in the late seventies, clients have been less and less able to afford endless models, and so rendering skills are, once again, becoming more important. This trend will probably continue because of the huge diversification of product lines to suit smaller and smaller market segments; instead of one portable radio, companies will produce several versions, each aimed at a particular type of person with his, or her, own lifestyle. This diversification will tend to re-emphasize styling (or 'appearance design') skills which will, in turn, bring drawing and rendering skills back into focus.

Above: *Livery for British Rail Freightliner container lorry by Don Tustin (1963). Gouache and brush onto board, with the container itself and the slight screen reflection airbrushed.*

Opposite page, bottom right: *Electric toaster (1957) – designer unknown. Water-colour on board with airbrushed background.*

2 Materials

There is a bewildering choice of drawing materials available and every designer has his or her preferences for this or that particular product. Having established that one works well and suits his or her particular style, a designer will stay with it and avoid picking up an alternative brand which might introduce a new, unforeseen effect at the wrong moment. Seasoned designers will need no advice on what to use, but for the newcomer it is often difficult, especially with a limited budget, to know what to buy.

It seems obvious to say so, but do buy the best available and learn how to look after your materials. Too many students, faced with a low budget, buy second-rate products which make an already difficult task even harder. What follows is a personal appraisal of currently available materials which I use, how they work and what to look out for.

Pencils

Pencils are the most commonly used drawing tool and have been for centuries. They are flexible, versatile and easy to rub out, and there is one for every kind of drawing. Some, like propelling pencils, were developed for drafting work but are also used for sketching. For working up ideas and sketching I prefer to use coloured pencils (see p. 17), despite their drawback of being difficult to erase.

Traditional pencils

Traditional wooden pencils are available in a range of hardnesses – nineteen in all – from 9H (very hard) to 8B (very soft). They are excellent for freehand sketching but tedious to keep sharp. I only ever use them for drawing from life because they are more sensitive to changes of pressure and angle.

Clutch pencils

Clutch pencils accept a range of leads in different degrees of hardness and in different colours. Pushing the button on the top releases the lead and allows it to be drawn out as the lead is used up and withdrawn into the barrel when not in use. A special sharpener is needed, so these have some of the disadvantages of traditional wooden pencils without their advantages of light weight and sensitivity.

Propelling pencils

Propelling pencils are similar to clutch pencils but with very fine leads available in four sizes (0.3, 0.5, 0.7 and 0.9mm) and several degrees of hardness. The leads are fine enough to obviate the need for sharpening. For most drawing purposes avoid heavy-duty or status-symbol pencils with self-propelling and auto-retract features and go for a lightweight construction with a smooth round profile; this makes them easy to revolve on long lines to obtain an even finish. I keep a boxed set of 0.5mm pencils with seven degrees of hardness 2B, B, HB, 2H, 3H, 4H, which are indispensable for working up a perspective from sketch through to finished drawing.

Traditional wooden pencils (right) are hard to beat for flexibility of line and shading but, of course, need constant sharpening. Many designers sketch with propelling pencils (left) although these are really more suitable for drafting. In the background are the Edding 'set of seven' pencils (together with refills), which are best kept together in their box. They come in a range of hardnesses from 2B to 4H and are light and a pleasure to use.

Papers and Boards

Layout paper

The designer's stock in trade, layout paper, is available in A-size pads and rolls in a selection of weights and types. Layout paper is ideal for setting up a perspective drawing because the drawing can be built up using underlays. The paper's slight opacity cuts out unwanted construction lines and yet is transparent enough for tracing through; it is also ideal for use with markers. The lighter the paper, the flatter the finish that can be obtained with the marker and unfortunately, unless coated, the more sheets under the top sheet will be ruined. To prevent this, some layout papers, like Letrapads, are coated on the underside to prevent bleeding through. These are sufficiently translucent to allow tracing through but give much greater brilliance because the colour is retained in the top sheet rather than being shared between two or three. On the negative side, their surface peels off very easily under masking tape and can sometimes break up under repeated marker application. With this type of pad you must also be careful, if you tear a sheet out, not to use the reverse side, as the coating makes it useless for working on.

Your choice of paper will depend on the type of finish you want; some designers prefer a lot of bleed and others like a tight finish. Surprisingly, there are enormous differences between brands; some are better than others and some respond better to one brand of marker than another. I keep two types of pad for everyday use: one (Frisk Studio 45 Layout Pad), which is lightweight and translucent, for working up ideas and underlays, and the other (Frisk Studio 60 Presentation Pad), which is slightly heavier and whiter but still translucent enough to see through, for finished drawings.

Coloured paper

Much depends on the type of drawing but I nearly always use Ingres paper because it accepts marker as well as pastel and crayon and is available in a wide range of colours. Like all fairly thick papers it absorbs marker quickly, which makes it difficult to obtain a flat finish. Ingres paper usually has a smooth finish on one side and a slightly corrugated finish on the other; unless you want a rougher texture, always use the smooth side.

Boards

For airbrush work I prefer a heavy (six-sheet) board with a smooth, hard surface which will not be disturbed by the removal of adhesive masks and will also allow mistakes to be scratched out. This type of hard-surface board, such as Frisk CS10, is expensive but stands up to a lot of abuse. Because it is comparatively non-absorbent, care is needed to avoid excessive build-up of ink which could eventually lift off at an awkward moment.

For general mounting of finished work be sure to get a board which is white, as many papers are semi-transparent and need the added intensity of a white background. They should also be rigid enough to stop the drawing bubbling off as the board flexes. I use a foamcore board (KAPA Featherlight) which has two surface sheets and a foam infill. It is therefore rigid and extremely light; this is very important when you are running for a train with thirty A2 concept boards tucked under your arm. Also, surprisingly, the sheer volume of thirty drawings when mounted to foamcore can do much to 'pad out' a thin presentation!

Rubbers

For nearly all occasions use a soft, plastic rubber which can be easily cut to obtain a clean crisp edge. Most are sold with a protective sleeve to keep the unused portion clean. I also keep a hard plastic rubber which holds its shape for longer when trimmed to a point or edge. For ink work on tracing paper or film, the new ink rubbers, which 'dissolve' the ink, work very well and can also do double duty as an everyday hard rubber.

Markers

The humble marker has come a long way since its introduction in the 1960s. There are now literally hundreds of different types for marking on every conceivable surface from paper to concrete and in line widths from 0.1mm to 40mm. From the designer's point of view the most important are 'art' or 'studio' markers, so called because they are available in a wide range of colours (over one hundred) and are designed specifically for studio use. Their development spawned a new type of presentation drawing – 'marker rendering'.

Most designers are fiercely defensive about their chosen brand of marker but few articulate the reasons why. The following is therefore a list of characteristics to look for when investing in markers:

Colour range. There should be at least a hundred colours in the range including a graded selection of 'warm' and 'cool' greys. Successful systems of the future will probably extend this feature to include graded tints of other colours as well. For example, all the primary and secondary colours might be available in a light and medium tint as well as full strength.

Colour consistency. If a colour starts to run out, you should be able to pick up another one and continue with no detectable change to the colour. Equally, if you buy a colour in London it should be indistinguishable from the same one bought in Edinburgh.

Colour continuity. When filling in a large area there should be no change to the colour across the work.

Nib shape and consistency. The most versatile nib is chisel-shaped for drawing both fine and broad lines. It should be hard enough to resist abrasion and maintain a crisp finish, yet soft enough to produce clean, flat colour. Nibs that are too hard produce a streaked effect which is difficult to lose, while nibs which are too soft tend to bleed and spatter. Some manufacturers, such as Pantone and Mecanorma, produce a selected range of colours in fine tips as well as broad, and one, AD Marker, offers four interchangeable nibs for maximum flexibility at the drawing board. This is a good idea in theory (it also saves money) but can be time-consuming and messy in practice.

Ease of use. Ergonomically, the marker should be easy to hold and work with, and to cap and un-cap efficiently. It should be clearly marked with a name (not just a number which can be difficult to memorize) and with a colour identification strip on the barrel rather than the cap. Most manufacturers try and match this colour strip to the contents, with varying degrees of success, so you should only use this as a rough guide and rely instead on your memory of that colour, or do a quick test on an adjacent piece of paper. The marker should be shaped to prevent it rolling off the drawing board and designed in such a way that, when a batch of them is in use, it is easy to select the right colour quickly.

Right: *A sequence of photographs showing the flexibility of line that the chisel-shaped tip of the marker offers – from quite fine, when used on its edge, to very broad, when used flat.*

Below right: *A selection of markers. The Holbein Illust marker range (from Japan) in the background is one of the most comprehensive available. On the left in the centre are a fine and a broad Pantone marker, flanked by two Design markers from Nouval in Japan – the one on the left is double-ended to guarantee colour continuity between the fine tip and bullet top. On the right are a Mecanorma marker, a Design marker (from the USA) and an Edding marker. In the centre are the original Magic Markers which are available worldwide and still one of the most popular with designers.*

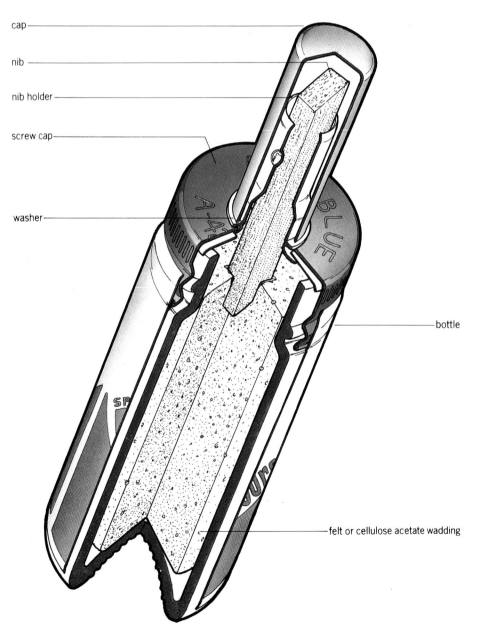

cap

nib

nib holder

screw cap

washer

bottle

felt or cellulose acetate wadding

Right: *This cut-away view shows all the parts that go to make up a Magic Marker. The manufacturing tolerances and quality control need to be very tight to prevent loss through evaporation and on-shelf deterioration. The felt wadding can be eased out with a scalpel and clipped into a bulldog clip for use as a giant marker.*

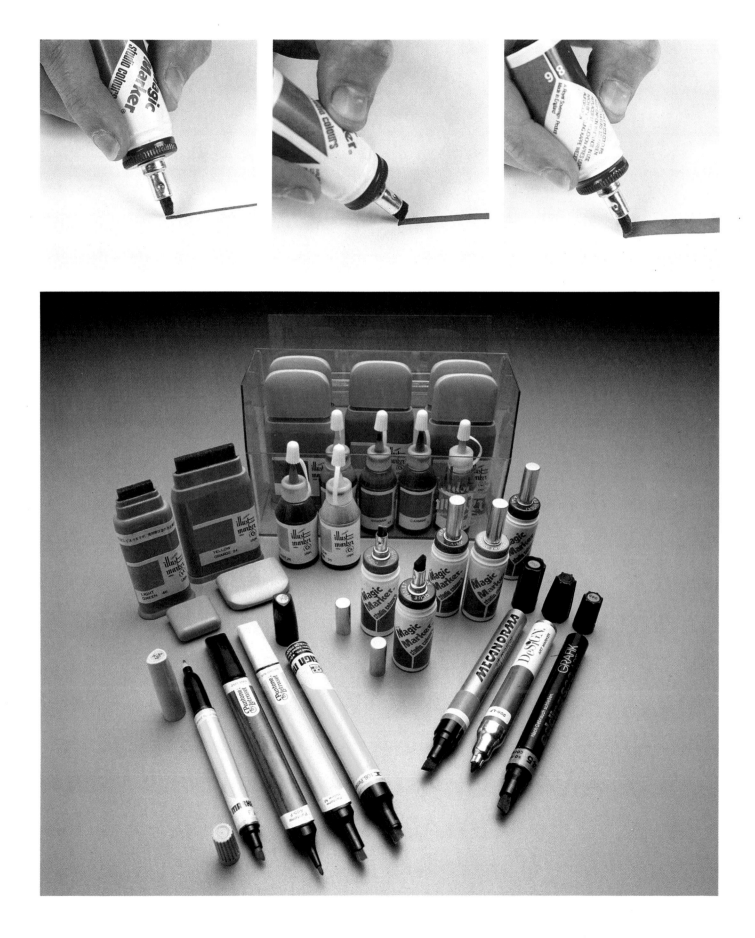

Flow characteristics

Markers should perform well in four main areas:

Bleeding. It should be possible to marker accurately up to a line without having to make allowances for the marker. Also, when working adjacent to a previously laid colour, there should be no bleeding of one colour into the next.

Colour pull. You should be able to work confidently with a light-coloured marker over the top of a well-dried darker colour without the light marker dissolving the dark colour and 'pulling' it into the new area of colour. It is a characteristic of all markers that they will do this if worked hard, but some brands are much worse than others.

Splash and blodge. This is caused by a loosely packed felt nib or an overfilled marker. If you pick up a new marker and work it to a line, there should be no splashing of colour across the line. The nib should be tight enough to avoid stray pieces coming loose, as this will cause unscheduled blodges.

Flat colour. It should be possible to infill an area with colour and leave it totally flat. The technique required to do this is described in more detail on page 61, but involves working fast enough to maintain a 'wet front'. This means constantly backtracking across previously covered areas to keep them wet and working the front edge of colour along until the area is filled. Obviously this is difficult to do with very large areas of colour and for these it is best to make your own 'giant' marker. This can be done either by wrapping cotton wool and lint around a piece of card, as shown in the picture, or by removing the wadding from the barrel and clipping it into a bulldog clip. Some areas that require infilling have such a convoluted shape that it is impossible to maintain a 'wet front'. In these instances be sure to leave the edge somewhere where it can be disguised later on, such as a highlight or 'shut-line'. (Shut-line is the industry term for the joint-line between two panels, one of which is usually an opening panel.)

Colour fastness

It would be nice to preserve marker work for posterity, but none of the ranges are resistant to ultra-violet degradation. It is, however, possible to extend the life of a marker drawing by spraying it with a UV-retardant fixative or by laminating it in plastic.

Above: *Going over a dry black area with a yellow marker should not cause colour 'pull'.*

Below: *A sequence of photos showing the construction of a giant marker using a piece of card, masking tape, cotton wool, a scalpel and some lint. Before applying ink, slightly wet the marker with cleanser/solvent.*

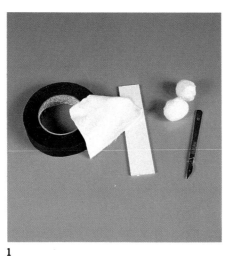

1

3

Papers

It is vital that you work on a paper or ground that suits your style and your markers. Not all papers are the same, and because the main method of drying is absorption and to a lesser extent evaporation, different papers will give different results. If you want a splodgy edge and large areas of flat semi-transparent colour, then use a lightweight layout paper; but if you want a clear crisp finish with dense colours, then go for a special marker pad.

Starting out

For someone buying markers for the first time it is difficult to know what colours to buy. Greys are available in two ranges, 'warm' and 'cool', consisting of nine shades which are numbered 1 to 9, from light to dark. (In this book marker colour names spelt with an initial capital letter indicate the Magic Marker range.) I advise the initial purchase of four Warm and four Cool Greys (perhaps Nos 2, 4, 6 and 8), a black and some primary colours. Thereafter the range can be extended as required but it is best to concentrate on those colours which extend the tonal range of the

2

4

primary colours. For example, a Cadmium Red might be supplemented by a Venetian Red to darken it down, a Cadmium Yellow by a Yellow Ochre, and a Mid Blue by an Antwerp and a Prussian Blue. Too many beginners use the range of greys to tone down colours. Avoid this if you can, because it usually makes the colour appear muddy.

Colour charts
It is important to keep a record of colours, preferably on the type of paper you use. This is essential for colour selection because it is an accurate record if you later need to replace a colour, or if you need a colour you do not use very often and therefore cannot remember its name. The charts issued by the manufacturers rarely resemble the actual colours.

Looking after markers
Be sure to recap markers tightly to avoid drying out. If a marker appears to be dry, this can often be the nib rather than the complete contents. There are some commercially available solvents (see p. 24) which will enliven the nib, allowing capillary action to restart and causing ink to flow again. It is also possible to use solvents to revitalize a totally 'dead' marker. Those that unscrew can simply be topped up with solvent, those that don't unscrew must be filled via the nib with a syringe. This is a useful practice for students because the cost of art markers is painfully high, but the results are usually less than perfect and, more important, unpredictable. The colour, much weaker than before, varies in intensity suddenly and often bleeds profusely; for these reasons keep topped up markers separately, and clearly identified.

A selection of the large range of coloured pencils available. The popular Berol Prismacolors (back right) are soft without being too waxy. In the foreground are the two Derwent ranges: the Studio set on the left is slightly harder than the Artist set on the right. The black barrels of the Studio set make it less easy to identify the colours at a glance. The Mitsubishi 'Uni' range (back left) are very similar to the Derwent Artist range.

Coloured Pencils

Choose pencils which are soft enough to obtain a soft tonal graduation, when held at a shallow angle to the paper, and yet hard enough to create a crisp line, without snapping, when held perpendicular to the paper. It is also important that they are available in a large range of colours (over sixty). Avoid waxy crayons which will smear and make subsequent overlaying with paint impossible.

Most manufacturers supply crayons in long flat boxes so that the full range may be displayed to advantage. These packs do not stand up too well in use and most designers end up using old tins and jars. Storing crayons 'point-up' in this way protects the point and makes colour selection easy; indeed, it would be nice if manufacturers supplied sets in specially designed tubs. If, like me, you store pencils in this way, you will avoid crayons which have only a colour-coded end, instead of a full coloured barrel. This makes them cheaper to manufacture but impossible to store, as the point breaks if

you store them 'point-down', and the colour-coded end is out of sight if stored 'point-up'.

The two most favoured types are Berol's Prismacolors and Derwent Cumberlands.

Right: Using coloured pencils

1. Coloured pencils should be hard enough to give a good tight line when sharpened and used upright; too soft a pencil will snap suddenly and leave a nasty mark on the drawing. For long lines always revolve the pencil to keep the width even.

2. To get a smooth tonal graduation, lay the pencil right over so that it is at a shallow angle to the paper.

3. If you are trying to put in a tone up to a curved line, use your fingernail to stop the pencil at the end of each stroke, sliding your finger along the line as you work. This allows you to keep the pencil moving quickly across the paper, which is essential for soft tonal transitions. If the line is straight, use a ruler as a stop.

4. To blend pencil tones further, use lint or a cotton-wool swab soaked in solvent.

Coloured Pastels

Coloured pastels should be conté- rather than oil-based, and soft but not too crumbly. It really is worth buying the boxed sets as they will last a lifetime if looked after, and only require occasional replenishing of the most used colours. (Be sure that the brand you invest in is also available in single sticks as well as sets.) I use Faber-Castell Polychromos which are available in seventy-two colours, more than enough because shades can easily be mixed. Their square section prevents them rolling annoyingly off the drawing board and they come in a wooden box, protected from breakages by a bubble-pack liner. The other most popular brand among designers are Rembrandt pastels which come in a huge selection of colours but are much softer than the Polychromos. This can make them smudgier and more difficult to control.

Pastels are enormously versatile and can be used in many different ways and in conjunction with many different materials. They can be used directly on the paper or scraped into a fine dust which can be applied with cotton wool. They can also be dissolved in solvents and used like markers in broad, bold, brush strokes. Finally, a piece of work can be easily masked off for application and the pastel rubbed away with an eraser to create highlights.

1

2

3

4

Below: Using coloured pastels to obtain a smooth finish

1. Scrape a small pile of dust off the side of the stick with a scalpel. If you do not have the exact colour you need, several sticks can be used together and mixed up in a small pile to obtain the desired colour.

2 (Below). Sprinkle some talcum powder onto the paper and lightly rub it in with a cotton-wool pad.

1

2

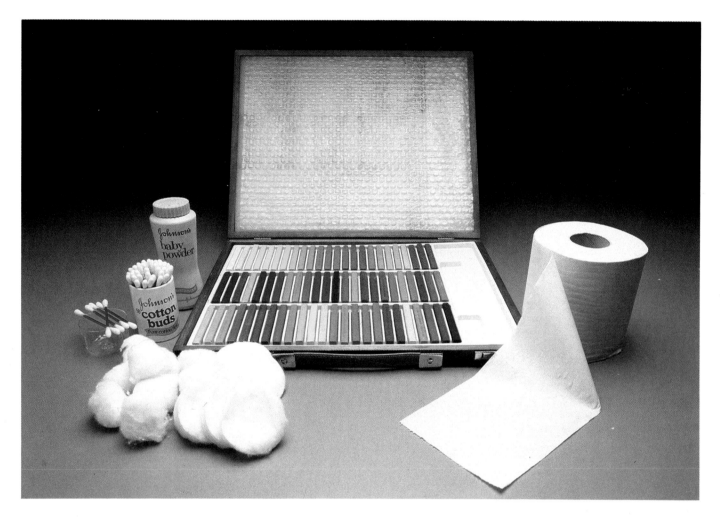

3 (Below). Taking some clean cotton wool, dab it in the pastel dust until you have sufficient loaded onto the pad, and apply it gently in smooth strokes. Don't worry about it spreading further than you want – concentrate on obtaining a clean finish. Remove excess dust with a toilet roll, and tidy the edges with a soft rubber.

Above: Faber-Castell's boxed set of seventy-two Polychromos pastels. As well as pastels you will need cotton wool (balls, pads and buds), talcum powder and a toilet roll.

4. (Below). Finally, if required, highlights can be created by rubbing away the pastel to reveal the white paper beneath.

Below: Using coloured pastels to obtain broad fields of colour
Lightly soak the pad with cleanser or solvent, dip it into a pile of pastel dust, and apply just like a marker. Unlike a marker, however, once the solvent has dried, the pastel can be rubbed out to create highlights (although not quite back to pure white paper).

3   4

Pens: Ballpoints, Felt Tips, etc.

There is a huge variety of 'pens' to choose from: felt tips, roller tips, fibre tips, ballpoints, brush pens, etc. with both water-soluble and permanent inks. Find out what suits you best but be careful with the soluble variety if you intend to use paint later, because the water dissolves the ink and discolours the paint. I keep fine- and medium-tipped permanent-ink pens in black for general edging work and sketching. It is important to choose one which is undisturbed by subsequent marker application and just visible under a black marker.

I rarely use coloured fineliners (like those available from the marker manufacturers) because I have yet to see them available in sufficient colours and with an ink/nib combination that gives an indistinguishable colour finish from their broader-tipped brothers. The best solution is probably that used in the Japanese Nouval marker which has a broad tip at one end and a fineliner at the other, both sharing the same ink and wadding – thus ensuring colour consistency. Many designers like to sketch with traditional ballpoints. Of course, ballpoint lines cannot be erased, although I have known this to be a positive advantage with some designers who welcome the way the technique 'forces' their design thinking and makes for economical sketching. Black ballpoints are also very useful for laying in details and crisping up edges on a rendering (like shut-lines on vehicle doors).

A selection of fineline pens. Ball Pentel on the right gives a good range of line thicknesses when it is leaned over and is unaffected when overlaid with marker. The Nikko Finepoint is also a very good all-round sketching, drawing and writing pen.

Paint and Inks

Before the advent of markers, gouache rendering was pretty much the norm for presentation drawings but the raw speed of markers has almost completely killed off the technique. Gouache is, however, indispensable for producing highlights and putting in very fine details like lettering, logos, etc., where opacity is essential.

For airbrushing the choice is between paint, which is opaque, and ink, which is translucent. Gouache is ideal for the former, but with the latter there is a wide choice of media. These break down into three categories: water-based inks, water-soluble inks which are water-resistant when dry, and solvent-based inks. The first category, water-based inks, are more difficult to use because overspraying can dissolve earlier applications. The second category, water-soluble inks, enjoy the most widespread use, although most are shellac-based and therefore necessitate rinsing through with methylated spirits or other appropriate solvent. I use Royal Sovereign's Magic Color because it is a water-soluble ink which can be freely mixed but dries to a water-resistant finish and requires no rinsing through of the airbrush with spirits (although they do supply a cleanser for dissolving dried ink in the airbrush). The third category, solvent-based inks, are used extensively in the auto industry for spraying full-size renderings. They are not designed for spraying and care must be taken to ensure adequate ventilation to avoid a build-up of dangerous fumes which is both a health and fire hazard.

Brushes

I keep three finest-quality sable brushes (sizes 00, 1, 3), and one size-10 synthetic watercolour brush (for mixing and loading airbrush inks). The golden rule with brushes is buy the best and look after them; never lend them to anyone, never leave them point down in water and always rinse them through immediately after use.

Airbrushes

Decide what you want an airbrush for before you buy one (see Chapter 6). Few designers have the time to do full airbrush renderings but there are still some things which are difficult to do any other way (and an airbrush is also useful for spraying small models). Students often believe that an airbrush is the answer to their rendering problems and rush out and buy one, only to discover that it isn't. It is an expensive item, so make sure you really

need it and look for one which suits your needs and buy the best. Stick to double-action brushes (down for air, back for paint) with a reasonable size cup. If you intend to use it for large-scale tape drawings go for a large-capacity suction-fed device, otherwise opt for a gravity-fed type.

Keep your airbrush clean and rinse it through if you need to leave it for all but the shortest periods.

Ancillary equipment

You will also need an air source which will be either an air compressor or cans of air propellant. I wouldn't recommend old car tyres (unpredictable supply and dirty air) or a footpump/tank (too energetic!). Your final choice will depend on how much you use an airbrush and how much you want to spend.

If you are really interested in airbrushing, it is worth buying a book devoted exclusively to the subject and borrowing an airbrush to experiment with. This may be difficult as my final recommendation is that if you own one, don't lend it to anyone.

New developments

The latest generation of airbrush-like products are the 'spraymarkers'. These are a welcome addition that bridges the gap between the marker and the airbrush. The Letrajet spraymarker blows air across the tip of a fineline Pantone marker in much the same way as a traditional 'diffuser'. The results are somewhat similar, in that the spray is fairly coarse. Nonetheless it is a potentially useful product that eliminates the tiresome cleaning-out associated with airbrushes.

Another new development is the electrostatic Spraypen which produces a truly superb fine spray that even an airbrush would find hard to beat. Because each ink particle is electrostatically charged, it is actually attracted to the substrate (earth). Equally, the spray is deflected by items with an opposite electrostatic charge such as plastic rulers, acetates and so on, which can lead to unpredictable results. Experimentation will be necessary in the early days of this revolutionary new product but its potential advantages are obvious: no external air source, compatibility with marker colours, easy colour change-over with no cleaning, long battery life and so on. Ultimately, if it could be invested with the flexibility of an airbrush, it could all but displace the airbrush in studios around the world.

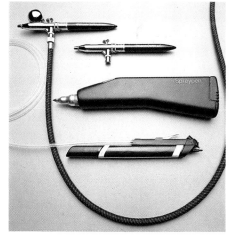

Above: This selection of paints and inks includes, in the background, a boxed set of Winsor and Newton's Designers' gouache – the best available. The brushes consist of a broad synthetic brush (left) and three sizes of finest-quality sable brush (right).

Far left: Airbrushes. On top are the classic De Vilbiss Aerograph Super 63 airbrushes – the A model (with the larger reservoir) and the E model – with, beneath, the prototype electrostatic Spraypen and the Letrajet.

Left: Air sources – a compressor and a can of propellant with its packaging. Choose a compressor that runs quietly, has a pressure tank and switches on and off automatically.

Technical Pens and Instruments

Most designers will already have a set of technical pens as they are indispensable for drafting work. There is little to choose between the various brands: each claims to be less prone to drying up or leakage than its rivals. New on the market are two types of technical pen working on a different principal. One, the Tombo PGS, uses a nylon tip which is replaceable and the other, the Pentel Ceranomatic, a ceramic tip. Both of these are very useful for general drawing work and don't clog so easily. On the negative side, they are slow to dry on non-absorbent surfaces like Polyester film and tracing paper, and this makes them frustrating to use for drafting work.

Technical instruments

Again most designers will already have a set of these. A useful addition for cutting circles, especially on masking film, is some compass blades, which slot in where the lead is usually clamped.

Scalpels, Tapes and Adhesives

Scalpels

There is an enormous choice of knives for every purpose but keeping one for this and one for that means keeping stocks of all the different types of blades. I use traditional surgical scalpels for practically everything; these have extremely sharp disposable blades in a variety of shapes. The handles are excellent for tasks requiring delicate control, like mask-cutting, but are rather less suitable for trimming out thick card. Rather than keeping a heavier-duty knife for this, I keep a second scalpel handle with the middle 'fleshed out' with masking tape so that more pressure can be applied.

Tapes

The most widely used tape is masking tape and I keep this in both a high tack (very sticky) and a low tack (less sticky). I also keep black photographic tape (in 2in, 1in, ¾in, ½in and ¼in sizes) for both masking and tape-drawing purposes (meaning, literally, drawing with tape – see p. 118). This is supplemented with rolls of super-fine, flexible black and white tapes, also used for tape drawing. Finally, it is useful to keep double-sided tapes and 'invisible' or 'frosted' tape for repairs to technical drawings.

Left: *Technical pens and instruments. In the background is a set of technical instruments and in the foreground technical pens with, on the left, Faber-Castell's TG 1 pens.*

Adhesives

I use an aerosol adhesive (Scotch Spray Mount by 3M) which is excellent for lighter, mounting jobs and especially useful for sticking down complex and delicate bits of paper. Gone are the days when a complex cut-out shape had to be carefully coated with gum. Spray Mount deposits a thin film of adhesive which won't squeeze out of the edges and is easy to reposition; bubbles and wrinkles can also be smoothed out. For heavier-duty mounting (i.e. paper on board) I use a stronger spray adhesive (Scotch Photo Mount by 3M). This doesn't have the repositioning capability but is less susceptible to bubbling off in changeable climatic conditions. (See p. 154.)

One word of caution: all these spray glues tend to hang in the air and are therefore easily inhaled into the lungs. This is a real health hazard so try and fix up an extractor to keep a constant flow of air to take the glue out of the studio. Some studios now ban the use of spray glues but this can seriously slow up work flow, so until someone comes up with a better idea make sure you minimize the risks.

Straight-Edges, Curves and Sweeps

Straight-edges

One of my most important and treasured pieces of equipment is a 600×70mm clear PVC straight-edge. This was guillotined from a 1mm thick sheet of flat PVC but could equally be made by scoring the sheet with a scriber or scalpel, snapping out the desired piece and then lightly finishing it with wet-and-dry paper.

The material is not attacked by most of the solvents found in markers and its whippiness makes it easy to use on a crowded board. Unlike a ruler, its transparency is unaffected by bevels and printing, and so allows a clear view of the work underneath. The larger-than-usual width is also important: when using a conventional ruler to set up perspectives, the edge not in use can distract the eye because it is too close to the edge in use; this can, surprisingly, make it more difficult to judge diminishing rates to an out-of-picture vanishing point. With this type of straight-edge there is a slight risk, particularly with technical pens, of the ink flooding under the edge; on these occasions lay a strip of masking tape on the underside to space the rule from the paper. Finally, never cut against your straight-edge; always use a steel rule.

Curves and sweeps

Clear PVC is also an ideal material for (radius) curves and sweeps. (A 'radius curve' is an

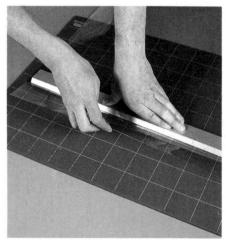

Making a straight-edge
You will need a sheet of clear PVC about 1mm thick. Score the surface with a scriber or scalpel to establish a line of stress (top), and then, leaning heavily on the ruler, snap upwards (above). Finish off the edge with wet-and-dry paper.

edge used to draw curves of constant radius, whereas a 'sweep' is an edge used to draw curves of changing radius.) Flexibility is essential when manoeuvring a long sweep on a crowded board, enabling one end to be used without shifting bits and pieces to make room for the other. Sets of curves and sweeps are difficult to find in shops, where they are often referred to as ship's curves and usually made of acrylic which is attacked by most solvents. They are quite expensive to buy but, like the straight-edge, easily made out of PVC sheet, provided you can borrow a set to trace around. This is done by carefully scribing around the master (being careful not to damage it); once a shallow groove is established the master is removed and the groove enlarged by further working with the

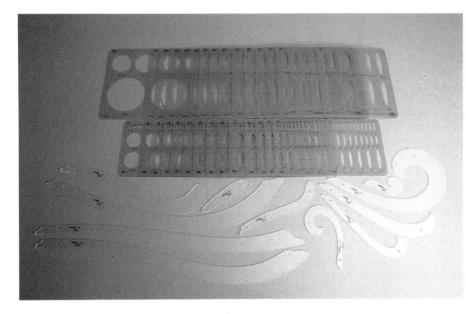

scriber or scalpel. Finally the curve can be snapped out with care, and finished with wet-and-dry paper.

Keep curves and straight-edges clean by wiping down with a rag, or tissue paper, soaked in solvent (check first that this doesn't dissolve the PVC). Doing this immediately after use with markers is an important habit to get into, particularly when changing colours. This prevents the reappearance of unwanted colour at a later date.

Ellipse Templates

Ellipse templates are very expensive and usually come in increments of 5 degrees, from 10 right up to 80 degrees, and in size (major axis), from 2mm to 250mm. They are typically available in small, large and extra-large sets. This is one occasion when it is worth investing in good-quality templates which should be thin, flexible and transparent, with one surface slightly matt to avoid reflection. They should be arranged so that the minor axes share a common line (preferably parallel to the edge of the template), and printed with fine lines to indicate both axes. Avoid the type that has ellipses segmented and arranged concentrically, as these are tedious and time-consuming to use, although, of course, they are unavoidable with extra-large ellipse templates. The best extra-large templates are like those made by the Alvin company, which allow you to see the whole of the ellipse. They are an application of the following property discovered in ellipse curves: the curve in any ellipse quadrant can be divided into two segments – the portion near the major axis having a curvature individual to that ellipse

only, while the portion near the minor axis possesses near-congruence to corresponding portions in the adjacent larger and smaller ellipses of the same degree.

It is possible to make the larger sizes of template in the same way as the curves and straight-edges, provided you can find someone who is prepared to lend you a set to work from (it's not a job for the faint-hearted). Use a thin material and don't try to make the smaller sizes.

Many students simply can't afford to invest in good-quality template sets and buy cheap substitutes that are more difficult to use. It is far better to buy good ones individually and build gradually to a full set. Unless you do a lot of technical illustration you won't need anything below 15 degrees (too thin) or above 60 degrees (getting close to a circle); start by buying a small 45-degree and a large 45. This may mean building a rendering around the ellipse guides available so, as soon as you can, buy a 30- and a 60-degree set and then gradually fill in the remaining increments starting with the 15.

Above: *In the background are Alvin (formerly Leitz) ellipse templates in large and small sizes; the extra-large ones are not shown. In the foreground are a selection of sweeps from Linex; these are acrylic which are more rigid than PVC but more vulnerable to attack by the solvents in markers.*

Miscellaneous

Board brush
A broad, soft brush is necessary for sweeping away rubbings from the work and from the drawing board. Don't use it for sweeping away excess pastel dust or it will transfer the colour to each successive drawing.

Bridge
A bridge is a rigid straight-edge, probably of acrylic, mounted at either end on blocks to keep it clear of the paper. This acts as a rest for the heel of the hand when using paint and minimizes the risk of smudging; it is also useful for drawing straight lines with a paint-brush. It is not, to my knowledge, commercially made but is easily put together with an old ruler or scrap of acrylic.

Talcum powder
Talcum powder 'lubricates' paper and ensures a smooth, easy application of pastel dust. (Looked at through a microscope, paper appears mountainous and full of troughs; talc simply fills in the troughs.) Talc can make erasing marginally more difficult because the eraser tends to slide rather than bite. Dusting with talc also knocks back the intensity of colour so that subsequent marker application creates a new tone. Finally, dusting with talc takes out the stickiness of marker 'pools' which have a tendency to form on low-absorption paper.

Carbon tetrachloride
This dissolves pencil lead, so is useful for cleaning pencil marks off tracings after the application of ink.

Solvents and solvent inks
Solvents are used for many purposes, such as cleaning the drawing-board and equipment, topping up markers, dissolving pastel dust, producing background effects, etc. The most widely used are Flo-master cleanser and inks, originally destined for use in refillable markers. Some of the more enlightened marker manufacturers make their solvents and inks commercially available, but these are often difficult to obtain because of their toxicity and flammability. It is a good idea to find out what the major constituent of the solvent in your favoured brand of marker is (probably xylene) and try and get hold of some. Failing that, use Flo-master. Lighter-fuel is just about acceptable for some brands of marker and is also indispensable for dissolving glues. Some of the solvents used in screen printing also work quite well. Experiment with what suits you and your brand of marker.

Tissue paper

A roll of soft toilet paper is ideal for keeping things clean, and is especially useful for removing excess pastel dust from the surface of a drawing without spreading or smudging it. Use the whole roll, revolving it and tearing off the stained sheets as you go.

Cotton wool, cotton pads and buds, and lint

These are useful for applying pastel dust and making your own broad markers.

Masking film

Masking film is essential for airbrushing, but check first that it won't pull away the surface you are working on.

Liquid mask

Liquid mask is a masking fluid that can be painted on, airbrushed over and then rubbed or peeled away; it is useful for fine work.

Tracing-down paper

Tracing-down paper is impregnated on one side with graphite or chalk and is used much like carbon paper for transferring a drawing onto another surface. It is available in graphite colouring, red, white, yellow and photographic 'drop-out' blue. Although they can be purchased in shops, it is easy to make a graphite sheet if you own a lead 'pointer' (sharpener). Sprinkle some solvent onto a cotton-wool or lint pad, dip it into the graphite shavings from the pointer and wipe evenly over some tracing or layout paper. Once dry, this makes a cheap and effective alternative to bought-in sheets, and cuts out the chore of 'back-coating' the underlay (see p. 102).

A selection of the other materials you will need. Clockwise from the left: board brush (from China), masking tape, Photo and Spray Mount, rotary pencil sharpener, battery-powered pencil sharpener, Flo-master ink and cleanser, Edding marker ink, black tape, bridge, pencil and ink rubbers, super thin black tapes from Letraset and Formaline, scalpels, a simple compass and a cutting mat.

3 Perspective Drawing

This chapter cannot hope to cover perspective drawing anywhere near as comprehensively as some of the specialist books on the subject. From the designer's point of view, however, much that is covered in these books is irrelevant. Many are aimed at architectural drawing which makes them difficult to adapt to the designer's needs, and most offer long-winded, if accurate, systems. This chapter therefore assumes a working knowledge of perspective and concentrates on cutting corners and producing reasonably accurate views quickly and easily.

Freehand Sketching

Good perspective is fundamental to good rendering. Without it the drawing has no chance of succeeding, and central to good perspective is basic drawing ability. This ability is not a manual dexterity, it is a cerebral process that hinges on the way we *see* things. Designers who draw well cannot only see immediately if a perspective is 'wrong' and how best to put it right but, more importantly, they are constantly developing their visual skills and honing their sensitivity to three-dimensional form. This command of three-dimensional form allows them to visualize and draw products as if transparent, and also enables them to organize internal parts quickly and easily into the optimum configuration. For example, the designer who can draw can more easily resolve complex mould shut-lines and the coring of injection-moulded parts.

There are two occasions when, despite considerable freehand drawing skills, a perspective system of some sort may become necessary. One is producing finished visuals from freehand sketches; many designers, although proficient enough, lack the confidence to do this and for them a perspective system can make all the difference. The other is when absolute accuracy is important, particularly if working from a GA, or General Arrangement, drawing (technical drawing that defines overall dimensions and layout). A good example of this would be a presentation where you want to show several alternative concepts all built around the same basic internals. Producing an accurate drawing of the internals, and using it as an underlay for the renderings, will ensure that each is in proportion to the next.

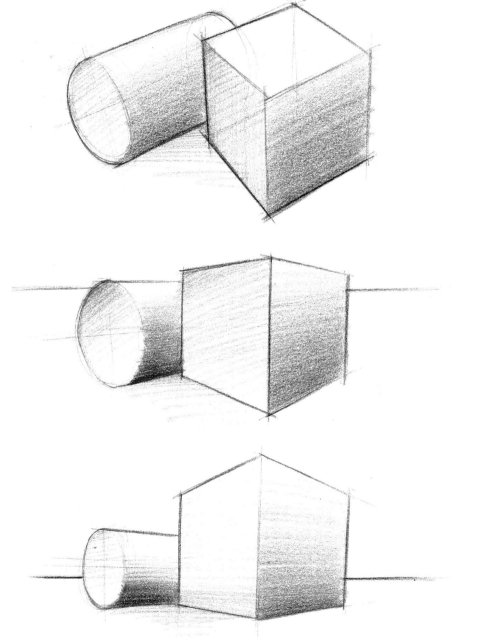

Three drawings to indicate scale
Three different views of the same two objects show how important it is to establish scale. This is dependent on the level of the horizon line, in other words whether you are looking down on, or up at, the object, and the rate of convergence of parallel lines, in other words how near, or far away, you are from it. In the first view the suggestion is of a small, perhaps table-top, object. The second suggests a bigger object such as a freight container, while the third is perhaps more appropriate to a building.

Choosing a View

The choice of view will depend on three factors. Firstly, it must show your design and its main features and details to their best advantage. Secondly, it must help define the scale of the product. This is determined by the position of the eyeline/horizon and the rate of convergence of parallel lines. Remember that small things are usually viewed from above and large things from lower down. Thirdly, it must be interesting to the viewer, which means that the composition of the whole drawing on the page will require careful attention.

There will be many occasions when you will want, and need, to disregard any, or all, of these three guidelines. This is particularly true if you want to get more 'drama' into the drawing by taking an ultra-low, or other unusual, viewpoint. Sometimes this can be achieved by simply repositioning the drawing on the page; indeed, many newcomers fail to appreciate that the perspective itself is still 'good' no matter which way up you position the drawing.

Size of Drawing

When I first left college I tended to do all my renderings far too small, to a point where they were actually becoming difficult to do. Because of, or as a result of, this, they also became oppressively tight and controlled. Later on, the average size grew as my confidence grew and I adopted a more fluid approach where my arm was doing the work rather than my wrist.

In our office we standardize presentations into A-size formats (with the exception of full-size renderings of large products). The final choice of size depends, to a degree, on the size of the product. Where possible, keep the drawing as near life size as you can without it dominating the page. This is clearly impractical for anything much bigger than a typewriter so, if your product is only marginally bigger, be sure to drop the size sufficiently to avoid confusion over scale.

The size of the drawing will also be influenced by your choice of media. For example, markers are not ideal for tackling tiny drawings and crayons are less suitable for large drawings.

Choosing a view
These three sketches of the same camera illustrate the kind of drawing that it is useful to do before embarking on a constructed perspective. In this case, either of the two upper views are sufficiently descriptive of the form and give a good impression of scale. The lower view might tell the viewer more about the detailing along the front but would not give a good impression of the whole. Remember to choose a view that shows the design and its key features to best advantage.

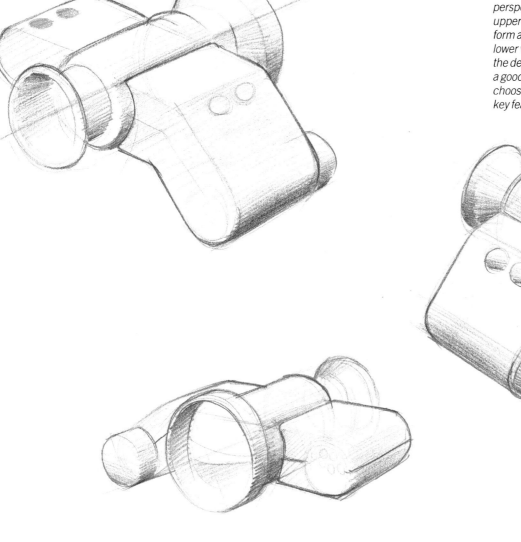

Setting up the Perspective

There are several shortcuts which you can take to avoid the chore of setting up a perspective from scratch. One is to use an existing photograph of a similar product, or take your own from the angle you require, and enlarge it, if necessary, to a suitable size. This may be achieved by either squaring it off (with a grid overlay) or by using a pantograph, Grant enlarger, PMT machine, Episcope, or photocopier. Another shortcut is to use sketch models, which you may well be working with anyway during the design process to help you evaluate ideas in three dimensions. Either photograph them or simply draw them from life.

If none of these methods is available, or suitable, then you will have to construct the drawing from scratch using a perspective system. The following pages outline some of the methods I use.

The Cube Method

I have found the easiest, simplest and most flexible method is to draw a cube in the view required for the drawing, because this is the basic building block of perspective construction and can also be used as the unit of measurement. In this way everything is considered in proportion rather than directly measured; a dimension is assessed as being twice that of another, rather than being, say, 50mm. Equally you might think of a dimension as being 2.5 cubes (i.e. units) in length.

If you use this system a lot, you will gradually build up a library of cubes and grids from many different viewpoints and you will simply reach into the drawer and select the right one.

For those starting from scratch, however, you will have to construct the basic cube.

Drawing the basic cube

There are a number of ways this can be done:

1. Actually make a cube out of card or plastic and photograph it from every conceivable viewpoint. Be sure to divide each face into four smaller squares, put in the diagonals and finally inscribe a circle, or series of concentric circles, on each side.
2. Use an off-the-shelf perspective grid.
3. Use a computer-generated cube. With the proliferation of micros it won't be too long before comprehensive perspective-generating programs are available to all. Even the smaller micros can handle simple programs to generate cubes and these can be revolved or zoomed in on at will, and then

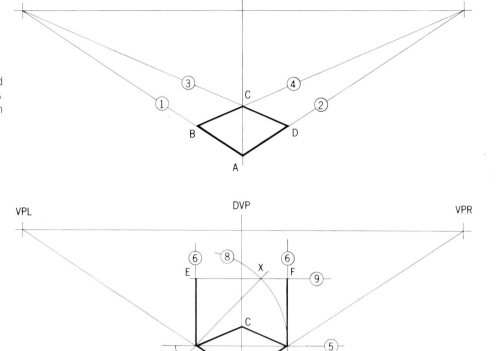

printed out for use as an underlay.
4. Construct a cube from an ellipse template, half by eye and half by construction (see p. 34).
5. Construct a cube using a traditional perspective system.

By now it cannot have escaped your attention that I have left the traditional method until last. There are many books about perspective and most designers are familiar with at least one of the mechanical methods of setting up a perspective drawing. Few designers have the time, or inclination, to do laboured perspectives. For this reason choose a method that is simple, such as that described by Jay Doblin in his excellent book

Perspective: A New System For Designers. The adjoining diagrams illustrate the construction of a 45/45-degree cube and a 30/60-degree cube to get you started.

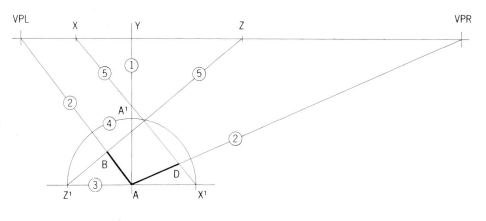

Left: Constructing a cube in 45/45-degree perspective

1.   Draw the horizon and position the two vanishing points (VPL and VPR) on it. Bisect the distance between the vanishing points to find the diagonal vanishing point (DVP). Drop a vertical from this point and draw two lines ① and ② from VPL and VPR at the desired angle A, which will be the angle at the base of the cube nearest to the viewer. Draw two more lines ③ and ④ from the vanishing points to intersect on the vertical. This forms the perspective square ABCD with the vertical AC as a diagonal.

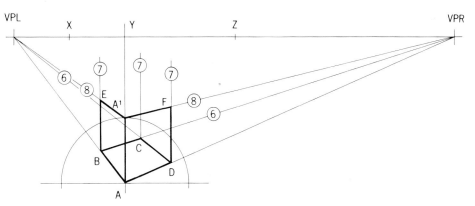

2.   Put in the other diagonal ⑤ which should be parallel to the horizon. Erect verticals from B and D ⑥. Place your compass on point B and swing an arc from point D through 45 degrees (⑦ and ⑧) to locate point X. Draw a horizontal line ⑨ through X to locate E and F, which completes the diagonal plane EFDB.

3.   Complete the top face of the cube by drawing lines ⑩ from both vanishing points through points E and F.

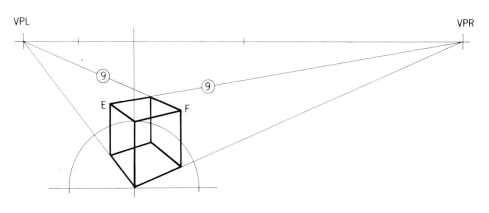

Above right: Constructing a cube in 30/60-degree perspective

1.   Draw the horizon and position the two vanishing points (VPL and VPR) on it. Bisect the distance between the vanishing points to locate Z. Bisect the line VPL/Z to find Y. Bisect the line VPL/Y to find X. Drop a vertical ① from Y and mark a point A at the desired distance below the horizon. Draw lines ② to point A to form the nearest angle of the cube. Draw a horizontal line ③ through A. Decide the height of the cube at A^1. Place your compass on point A and draw an arc ④ through A^1 to locate points Z^1 and X^1 on the horizontal line ③. Draw lines X/X^1 and Z/Z^1 ⑤ and locate their intersection points on lines ② at B and D.

2.   Draw lines ⑥ to complete the perspective square ABCD. Erect the remaining three verticals ⑦. Draw lines ⑧ to A^1 to locate E and F.

3.   Complete the cube by drawing perspective lines ⑨ to E and F.

Subdividing and extending the cube

The diagrams on this page and the next show
first how to divide up the cube into smaller
units by drawing diagonals and then how to
extend it into a matrix in all three directions. It
is most important as you extend the original
cube to constantly cross-check what you are
doing and avoid the build-up of error. As you
project a new point into space, double-check
it by independently projecting from another
source, such as a diagonal. You will quickly
learn that distortion is inevitable. Swing a
circle from a point which is mid-way between
the vanishing points and passes through
them. As your image approaches this circle,
the level of distortion will become increasingly
unacceptable.

Surprisingly, many students fall into the
trap of making the nearest corner of their
cube, or grid, less than 90 degrees. There are
no circumstances when this can be possible.
Try holding a record cover at eye level, it will
appear as a line; now put it on the floor and
look at it in plan-view, all the corners are true
right angles. As you look at it in every position
between these two, the nearest angle will vary
from 180 to 90 degrees; it can never be less.
Any cube whose nearest corner touches the
circle drawn through the vanishing points
must have a nearest angle of 90 degrees (two
lines drawn from the points of intersection of
the diameter and circumference to any point
on the circumference must meet at right
angles) and must therefore be distorted.

The golden rule when extending your cube,
or doing any kind of perspective, is to trust
your eye and not the system. You can help
yourself in this respect by adopting a different
viewpoint to the picture from time to time.
There are several ways of doing this:

1. Screw up your eyes to cut out distracting
construction lines.
2. Step back from your drawing board
frequently, so that you can take in the whole
drawing without having to move your eye.
3. Look through a reducing glass, which
gives the same effect as stepping back.
4. If the paper is transparent enough, turn it
over and look from the other side. This is the
most useful method because, having spotted
an error, you can correct it on the back. Turn
the paper over again and erase the old lines to
reveal the new ones showing through from
behind. If you are working on opaque paper
you can achieve the same effect of viewing
from behind by looking at it in a mirror.

Your eye can quickly come to accept an error
which can gradually get out of control, so
make at least one of the above checks
frequently to keep everything in shape.

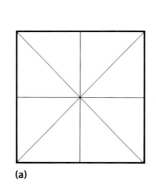

(a)

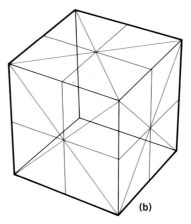

(b)

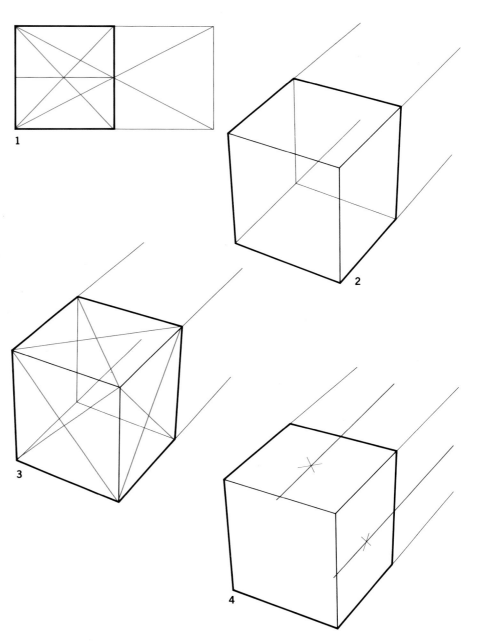

1

2

3

4

Left: Subdividing the cube

(a) As with any perspective situation, look first at a measured elevation. This information is then easily transferred to the perspective view.

(b) The cube can therefore be subdivided by drawing diagonals to find the mid-point of each face and then dividing it into four smaller squares. For the inexperienced eye, this is not as easy as it looks. Each vertical and horizontal line must converge at the same rate in order to appear parallel. Help yourself to do this by comparing them with the hidden lines of the cube which are nearer and therefore easier to judge from.

Left and right: Extending the cube into a matrix

1. As with subdividing the cube, look first at an elevation. You can see that a single square can be extended into a rectangle by finding the mid-point of one side and drawing diagonal lines from each corner through this point to find the opposite corner of the rectangle. This information can easily be transferred to the perspective view.

2. Extend the 'horizontals' in the desired direction towards an imaginary vanishing point.

3. Draw the diagonals to find the mid-points of each face of the cube.

4. Draw perspective lines through these centres to find the mid-points of each opposite line.

5. Draw diagonal lines through these points (and remember that these are the diagonals of the perspective rectangle) to locate the corners of the new cube.

6. Complete the new cube by connecting up these corners. As you do so, check that the rate of convergence is consistent with adjacent lines.

7. When extending a cube into a matrix, be sure to constantly cross-reference by also drawing the diagonals of the new, larger square as well as using the diagonals of the rectangle.

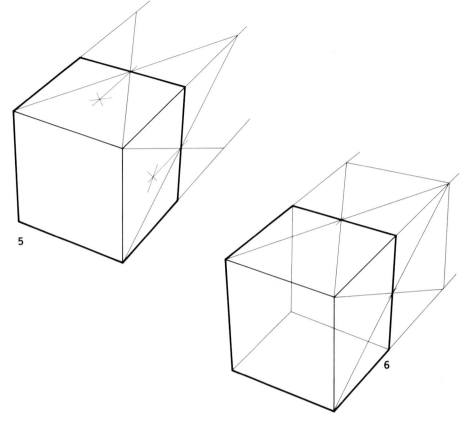

5

6

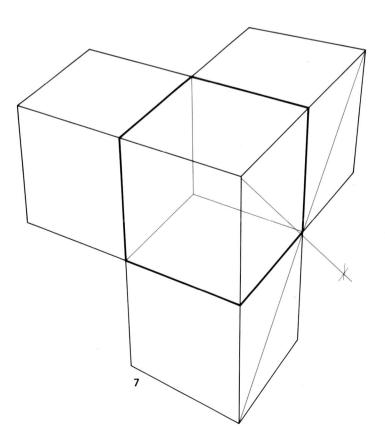

7

Ellipses and Circles in Perspective

To help understand a circle in perspective, look first at a plan-view of a circle circumscribed by a square (as illustrated opposite page, top). Notice that the centre of the circle coincides with the centre of the square and that the circle touches the square at the mid-point of each side. It should therefore be a simple matter to transpose these conditions onto a square drawn in perspective.

It can be shown mathematically that a circle in perspective is a true geometric ellipse, so it will also help to look at a true ellipse to see whether there is any help to be had from this end. An ellipse has a major and a minor axis at right angles to each other, it is also symmetrical about both axes so that each half of each axis is the same length. When an ellipse is viewed as a circle in perspective, the axis of rotation of that circle coincides with the minor axis.

This is the most common mistake made by students when first attempting to draw circles in perspective (except, strangely, those parallel to the ground like cups), because regardless of the view of the circle they nearly always draw the ellipse with the major axis vertical. If you are in any doubt about this, try drawing a range of perspective cubes above, below and on the horizon. If you then draw in an ellipse on one of the vertical faces, only on the horizon is the ellipse upright, and the more the cube is above or below the horizon the more the ellipse is leaned over. If the circle is parallel to the ground, it can be seen that its axis of rotation will be vertically downwards, and therefore the minor axis of the ellipse will also be vertically downwards, with the major axis at right angles to it and therefore parallel with the horizon. This is usually the easiest way to draw ellipses so, if you have trouble drawing freehand ellipses, turn the paper so that the minor axis is vertical on the paper and the major axis horizontal; it is also helpful to tick off lightly the intended distances along the axis.

Rules for drawing circles in perspective are therefore:

1. A circle in perspective is a true geometric ellipse.
2. A circle in perspective will touch its circumscribing square at the mid-point of each side.
3. A circle in perspective and its circumscribing square will share the same centre.
4. The ellipse will be geometrically symmetrical about its axis.

5. The axis of rotation of the circle will coincide with the minor axis of the ellipse.

Although these are important rules for the designer it is only fair to say that they are not totally correct (see opposite page). For our purposes, however, they are entirely adequate.

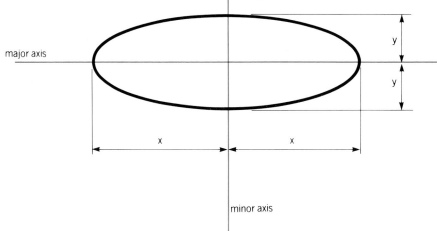

Above: Properties of ellipses
A geometric ellipse has a major and minor axis and is symmetrical about these axes (x=x and y=y).

Below: Axis of rotation
When an ellipse is viewed as a circle in perspective, the axis of rotation of that circle coincides with the minor axis. In other words, if the circle you are trying to draw was made of cardboard and you could spin it, then its axle, which of course must pass through the circle's centre and at right angles to it, will coincide with the minor axis of the ellipse.

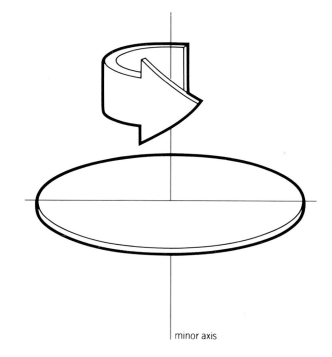

Right: Anomaly

1. As with the cube, look first at an elevation of the circle and its circumscribing square. The centre of the circle and the centre of the square coincide, and the circle touches the sides of the square at their mid-points.

2. This information can be transferred to the perspective view. However, as a student it seemed to me that a circle in perspective could not possibly be a true ellipse as can be shown thus:
Construct a square in perspective with its centre on the horizon. Next, divide it in half vertically by dropping a vertical line through the centre. (It will already be divided horizontally by the horizon.) As we have seen, our perspective circle will touch the circumscribing square at the mid-points of each side and will share the same perspective centre. Now, because of the effects of diminishing distance, that half of the square nearest to the vanishing point (x^1) must be smaller than that half further away from it (x). Equally, therefore, that half of the ellipse nearer the vanishing point must be smaller than that half nearer to you the viewer. Therefore it is not symmetrical and so cannot be a true geometric ellipse. The anomaly here is that, while a circle in perspective is indeed a true ellipse with its minor axis coinciding with the axis of rotation, the geometric centre of the ellipse does not coincide with the centre of the square *but is in fact shifted slightly towards the viewer.*

3. A schematic diagram may help you understand why this is so. Draw a circle with your eyepoint some way from it. Next, draw visual rays from this point, which are tangential to the circle, and then connect these tangent points together. What you actually see when looking at a circle in perspective is this chord, and the further away your eyepoint the nearer this centre gets to the true one.

4 and 5. For the same reason, it can be seen that the ellipses used to draw concentric circles in perspective cannot share the same geometric centre but will be offset slightly one from another. For all practical purposes this anomaly can be ignored because the offset is so small. Only those who are involved in super accurate measured construction and who will, in any case, be using a proper system need worry. Certainly your client will probably care little for the niceties of perspective theory!

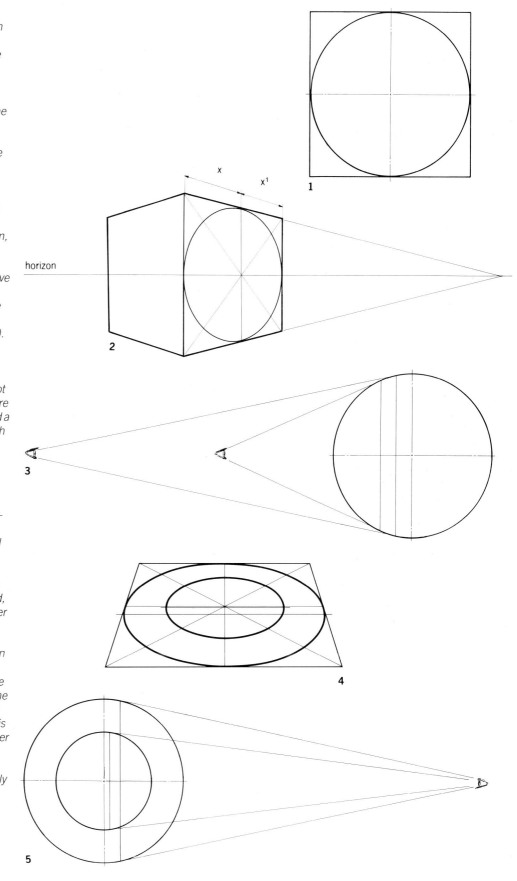

horizon

Ellipse guides/templates

Most designers will keep a set of ellipse guides simply because it is very hard to draw crisp ellipses freehand or even using French curves. Nevertheless, as with all perspective, do practise drawing ellipses freehand at all angles of inclination or you will find yourself depending on the guides instead of your eye. Remember that ellipse guides are just that – guides.

To draw an ellipse within a constructed circumscribing square, first find its centre using diagonals and then draw in the axis of rotation and a line at geometric right angles to it. Next choose the ellipse whose perimeter comes closest to touching the mid-point of each side of the square when the minor axis is aligned with the axis of rotation and the major axis is aligned with the line at right angles to it. You are unlikely, even with a full set of guides, to find the *exact* ellipse required in both size and inclination, so use the nearest one to it. If this still looks unsatisfactory, it is usually possible to move the guide slightly as you draw each half of the ellipse to approximate to the 'in-between' ellipses.

Remember also, that with a cylindrical object the ellipse nearest the viewer may be, say, a 35-degree ellipse but, as the cylinder recedes, the ellipses will come closer to the plan-view and you will find that a 40-degree and then a 45-degree will be a better fit in the circumscribing square. However, the further away the circumscribing square is from the viewer, the more distorted it will become and, equally, the more difficult it will be to accommodate an ellipse.

It is also useful to bear in mind, particularly for those with a limited selection of guides, that, if the product to be drawn is generally cylindrical, like a bottle or screwdriver, an ellipse template can be used to create the view without needing to start from scratch.

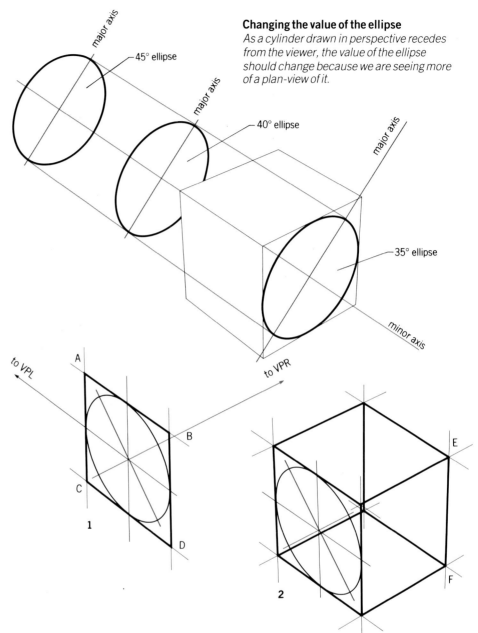

Changing the value of the ellipse

As a cylinder drawn in perspective recedes from the viewer, the value of the ellipse should change because we are seeing more of a plan-view of it.

Constructing views from an ellipse guide

1. Lay in an ellipse of the right size with the minor axis at approximately the right inclination for your intended view; be sure to mark off the major and minor axes. Extend the minor axis in both directions and remember that this line must go to one of the vanishing points (VPR). Next drop two verticals, AC and BD, which are tangent to the perimeter and then draw a line that runs through the two tangent points and the centre of the ellipse; this line must go to the other vanishing point (VPL). Drop another vertical through the centre of the ellipse.

So far, little skill or judgement has been called for. The next step is to finish off the circumscribing square and to do this you will need to estimate the rate of diminishment (and therefore the proximity of the vanishing point) of the remaining two lines, AB and CD, with respect to the centre line already established. The two lines should recede to the vanishing point and each should be equidistant from the centre line as measured along a vertical drawn at any point. Be careful not to impose too great a degree of diminishment or the vanishing point will be too close and therefore distortion will be more likely. The resulting trapezoid should be a circumscribing square in perspective (given the offset centre anomaly we have chosen to ignore).

2. Next, use your judgement to draw in four perspective lines from each corner of the square to the other vanishing point, bearing in mind that they should all converge at the same rate. Start with the line closest to the axis of rotation and work outwards. Obviously, if the vanishing points are not 'off the board' and you want more accuracy, you can put in a horizon from the previously established vanishing point (VPL) and where this horizon line intersects the axis of rotation will be the second vanishing point (VPR). Next, you will have to estimate where to put the vertical (EF), which defines the second side of the cube, and this you will have to do completely by eye. Do it lightly at first and complete the rest of the cube, especially the unseen faces, to give you the maximum visual information on which to base a decision. Look at the completed cube and decide whether it looks distorted. If it looks out of proportion, move the line to correct it and complete the rest of the cube again. With this method, remember that every step of the way, distortion can creep in, so it should only be treated as a quick guide.

Right: Constructing a circular shape using an ellipse guide

1. Ellipse guides can also be very useful for building up views of complex circular shapes, such as bottles, from elevations, provided their axis of rotation is vertical. Simply take the elevation (or a single profile and centre-line) and draw horizontal section lines at regular intervals and at key changes of direction.

2. Because the bottle is small, we can use a single template. If, on the other hand, it was a much larger object, we would select one for the top, one for the middle and one for the base. Draw the ellipse which, with both axes aligned, has the profile passing through the point of intersection of the major axis and the perimeter of the ellipse. This will produce a skeletal view, and it only remains to connect up the resulting silhouette to complete the construction. Note that the silhouette and cross-section diverge away from one another as the bottle flares outwards and then in again at the base.

3. The whole process is equally useful but less easy when drawing the same product on its side. In this case it is obviously necessary to first construct, or sketch, the elevational view in perspective complete with vertical section lines. If you need to set up an accurate perspective, then this method probably isn't worth the trouble because you may as well construct all the necessary circumscribing squares. If, however, you are reasonably confident that you can do this at least partly by eye, it will save a lot of time. (You need to be able to divide a single line into equal parts with a regular rate of diminishment. The way I do this is to draw the line and mark off each end; then I estimate the mid-point in perspective to one side of the measured centre, divide the two halves in half again and so on). Next, draw a geometric right angle through each section centre to indicate the major axis. Then, offer up the ellipse guides until you find one which, when both axes are aligned, has its circumference passing through the point of intersection of the section-line and its corresponding vertical. Repeat this for each section to create a skeletal view and then join up the silhouette.

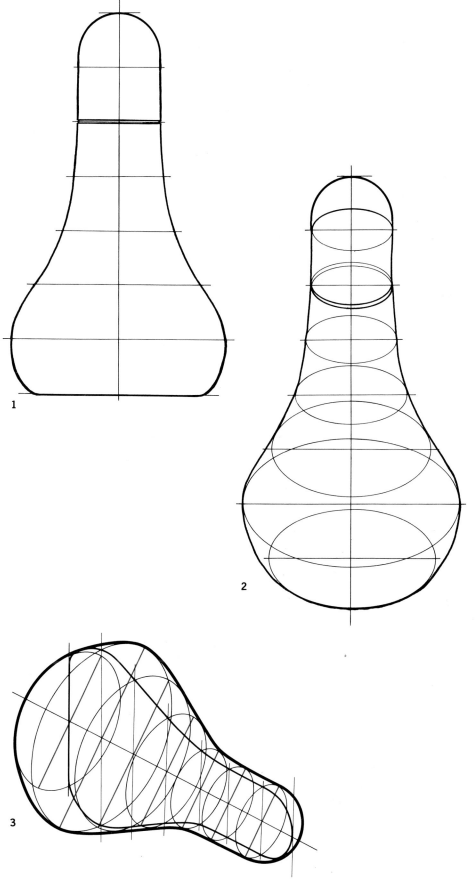

Dividing up a Circle in Perspective
There will be many occasions when you will have to divide up a circle in perspective into segments. With a circumscribing square this is fairly easy because you already have increments of 45 degrees established by the diagonals. Finer divisions can be obtained by using your eye to divide each segment in half again (or into thirds). Do this on opposite segments and check for accuracy by joining up through the centre.

If you are not working within a circumscribing square, use the major and minor axes as the established 90-degree angles and use your eye to subdivide into 45-degree sections and then in half again, checking across the centre as you go. As with dividing a line in perspective into equal parts, remember that this is not a geometrically equal division because the equal segments are not evenly disposed around the perimeter of the ellipse, but are concertinaed around the major axis. This is much more noticeable with the 'thin' ellipses. If accuracy is essential, many template manufacturers print a protractor around the largest ellipses with increments already marked off.

Spheres

A sphere in perspective is, of course, a perfect geometric circle and can therefore be drawn with a compass. Invariably, though, this is insufficient because more detail is required or because the spherical part must relate to another part of the product. For nearly every application it is necessary to draw the equatorial ellipse and at least two others; the diagrams below illustrate how to do this.

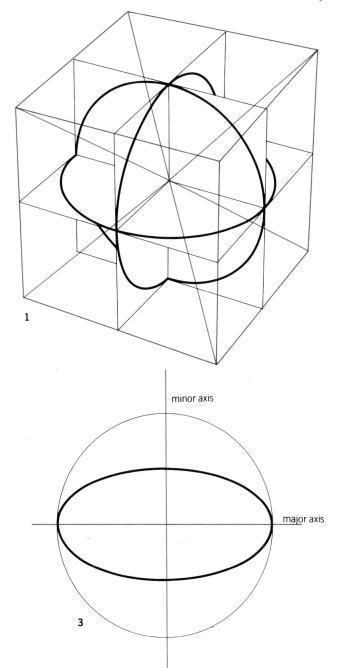

1

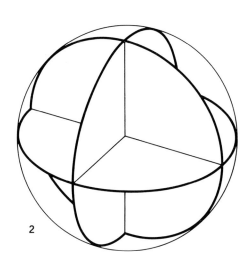

2

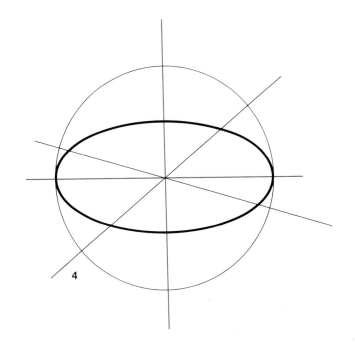

minor axis

major axis

3

4

Spheres

1 and 2.   A sphere in perspective is, of course, a true geometric circle. More often than not, however, you will need more information than is provided by simply drawing a circle. You will probably need the equator and any two ellipses which run through the north and south poles at right angles to each other. This can be done by first constructing a circumscribing cube and then working inwards by finding the centre of the cube using diagonals. Set up three circumscribing squares and their corresponding ellipses and then swing a circle of the appropriate radius to complete the sphere.

3.   If, however, your product is nearly all spherical, then it is possible to work backwards using ellipse templates. To do this, first choose the ellipse for the equator remembering that this will determine how much you will be looking down on the sphere. Draw in the ellipse and both its axes, and swing a circle, of the same diameter as the major axis, from the centre.

4.   Next, divide the ellipse into four equal segments either by eye or with an ellipse protractor. This creates two lines at perspective right-angles to each other.

5.   Draw a line passing through the centre and at geometric right-angles to one of these two lines. You now have the major and minor axes for an ellipse and two of the points, X, on the circumference of the first ellipse, through which it should pass. Find the new ellipse (with the same length major axis as the equatorial ellipse) that, when aligned with both axes, passes through both points X.

6.   Repeat the process for the other side.

7.   The intersection of these two ellipses on the original minor axis locates the north and south poles.

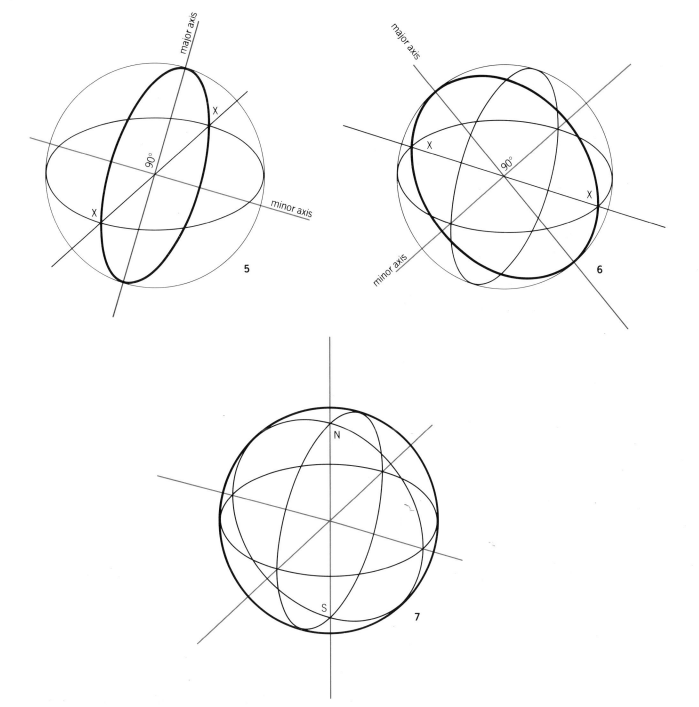

Radii

One of the most common characteristics of contemporary products is the radiussed edge. Radii help to soften a product and reduce its visual bulk; they are also predictable and easy to manufacture. It is important to understand how to draw them because you will then see exactly why they contribute to reducing visual bulk. It will also help you understand parts of the following chapter which describe how we perceive reflections in radiussed forms.

Radiussed edges

As an illustration of what is involved in constructing radiussed edges, the following diagrams show how to draw them onto the basic cube.

1.   Look first at the side elevation and observe where the centres of the radii and spheres lie in relation to the outside skin. The side of a radiussed cube is made up of square flat areas interconnected by cylinders at the edges, with the quadrants of a sphere in each corner. Armed with this information it is relatively simple to construct a cube, with spheres and cylinders at the appropriate places, working from the centre-line outwards.

Most of the time, however, the radii are not such a dominant feature of the product and therefore do not warrant such accurate construction.

2.   First of all, draw a cube and lay in the lines which describe the extremities of the radii, i.e. the lines at which the flat planes become rounded. These lines are crucial because they define the point at which the form changes direction and the point where light (and therefore colour) begins to change. If you analyse the nearside corner, itself a cube, you can see how this can be divided into three intersecting ellipses which themselves define a sphere. The minor axes of each ellipse form the back three lines of the cube. I find it helpful to visualize each ellipse quadrant as the line which marks the change of direction between that which is cylindrical and that which is spherical.

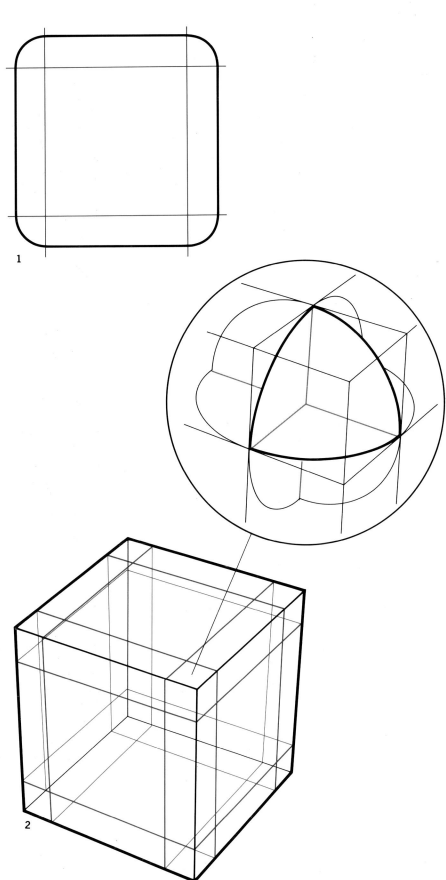

3. All that remains is to freehand in the ellipse quadrants on each corner. It helps, at this point, to think of the cube as three separate 'ribbons' wrapping around at 90 degrees to each other. Starting with the left nearside face, wrap one ribbon over onto the top by sketching in a smooth curve between the two faces. Next, wrap it over onto the rear face in the same way, drawing in first the two curves, and then the new falling-away edge tangent to them. Continue the wrap down the back, on to the base and back up the front so that you have a complete ribbon. Be sure to work as carefully on the unseen planes as on the front, at least until you have sufficient confidence.

4. Use exactly the same process starting with the right nearside face, working over the top and back across the base until you have another ribbon at right angles to the first.

5. Finally, work from the left nearside face across to the right nearside face, and then around the two unseen faces and back to the front.

6. All that remains to be drawn are the missing corners, which can easily be sketched in freehand. The final underlay has all the necessary information about changes of direction which we will need for later colouring up; note, too, how the radiussed cube looks so much smaller than the original because all the falling-away edges have moved inwards.

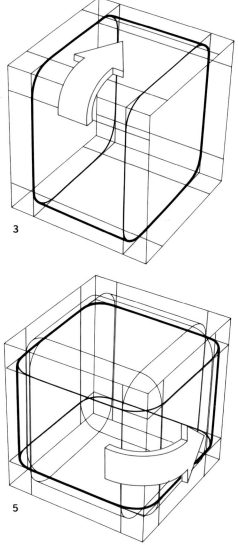
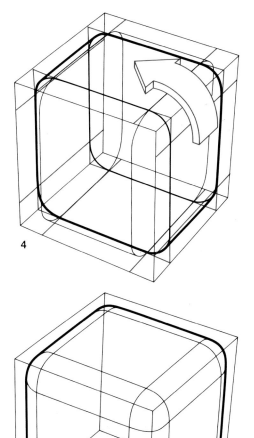

Compound Forms

Armed with the ability to draw rectilinear and circular forms in perspective (cubes, cylinders, cones and spheres), it is relatively easy to construct much more complex shapes by building up a perspective grid and working from side, front, and end elevations. Once a grid is established, the appropriate cross-sections can be drawn in perspective. These are usually the centre lines of the product, and you are therefore working from the inside to the outside of the product. Once sufficient sections have been drawn in, it is simple to complete the drawing freehand.

Three-Point Perspective

Most students taking their first steps into the world of perspective will want to begin with a two-point perspective where all verticals are parallel. To my eye, this always looks wrong because it does not correspond with reality. Remember that a good eye for perspective is essential for sketching and keeping everything in shape; and you cannot hope to educate your eye if you always work with a two-point system which ignores converging verticals. By all means start with a cube drawn in two-point perspective, but then move in the verticals as evenly as possible so that they lead approximately towards the third vanishing point. If you are unsure, quickly extend the cube to the required dimension and check that you haven't overexaggerated by imposing too much convergence in the verticals.

Conclusion

Examine the stage-by-stage examples and finished underlays on the following pages, so that you understand the process of building up a view. Then try it for yourself with one of your own designs.

It is pointless to move onto the following chapters if you are not confident that the underlay you have produced for your product is accurate. No amount of careful colouring up or jazzy backgrounds will conceal a distorted perspective. Go back and try to get it right first. With perspective, practice really does make perfect because you are constantly educating your eye; the more you do it, the better you will become.

Example 1:
From GA Drawing to Underlay

Hairdryer

The following example takes simple
elevational views from basic cube through to
perspective line drawing (master underlay).

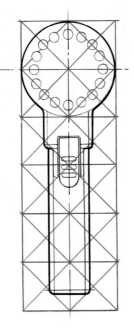

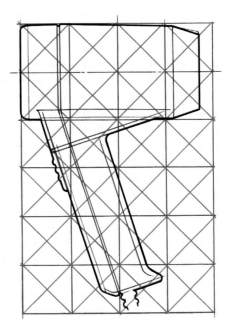

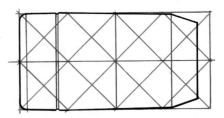

1

Stages

1. This is a fairly typical example of a GA
drawing that a designer might be faced with.
For technical reasons he or she may be
working from scale drawings to ensure that
the product will go together and that
everything is in proportion. The first thing to
do is to decide what view of the hairdryer is
required, in this case a rear three-quarter
view. This was sketched out freehand first to
check that the view described the form
adequately and to establish the most logical
starting point for constructing the grid. Since
the back, top edge of the dryer is nearest to
the viewer, the GA was squared up from this
point using the overall diameter of the circle
as the cube's dimension. Our grid will
therefore be approximately three cubes high,
two cubes long and one cube across.

2. The nearest cube is drawn first (in this
case using a cube left over from an earlier
drawing). This is then extended one cube to
the right and three down, checking all the
time that distortion is under control.

3. The side elevation of the dryer is then
sketched in along the centre line of the grid.
The major ellipses are also put in lightly, using
guides. Because the handle is suspended
above the base of the grid, it is easier to
project the lines which describe the edges of
the handle to the bottom of the original GA
grid and then mark those points onto the base
of the perspective matrix; these can then be
projected back up. In this way the angle of the
handle can be accurately determined.

4. At this point, to avoid confusion, the matrix
was slid under another sheet of layout paper
to cut out unwanted construction lines. The
elevational section can be extended either
side to the required width, and the lines which
define the intersection of handle and cylinder
become clearer. The cylindrical lines are now
put in more clearly, particularly the two which
define the radius around the back rim.

5. The radii along the handle are first defined
(in exactly the same way as we did with the
cube) to establish how far the falling-away
edges must be moved in. The rear face of the
dryer is divided first into four, then eight, then
sixteen to establish the centres for the intake
holes. These can then be put in easily, using
the same value of ellipse guide as for the
larger circle but in a much smaller size. The
other details such as the switch and grommet
are put in by eye last of all.

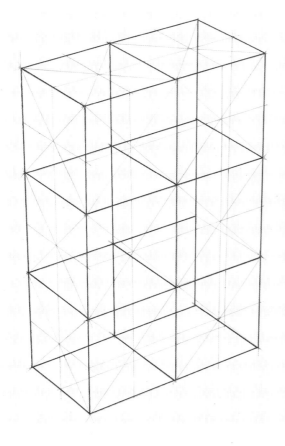

2

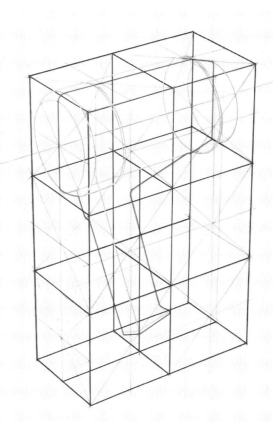

3

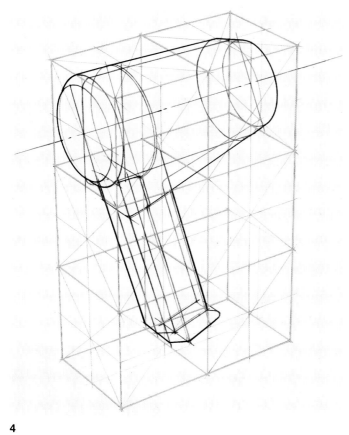

4

5

Example 2:

Freehand Underlay

Camera

This example is very similar to the hairdryer but is done freehand without a grid.

Stages

1. The major cylindrical form is drawn first. This is done with a 45-degree template which is stepped along the minor axis of the cylinder to produce the lens, main body and viewfinder. As you work along the minor axis, make sure you raise a vertical within each ellipse to keep the sectional view correct. The unit which hinges off the main cylinder is projected using the major axis of ellipses already used for the main body. Verticals are also dropped tangential to the lens hood and their centres joined up to establish a horizontal and approximate left-hand vanishing point. The major vertical down the face of the lens hood is also extended as the main guide for locating the front of the handle. As with the hairdryer, once the centre-line of the handle is established, it can be extended to left and right to define the handle width before the radii are put in.

2. All the main parts are firmed up and the details lightly laid in. The microphone cover is constructed by dividing the front ellipse into the right size of segments and wrapping the

divisions around onto the cylinder. Concentric ellipses are then put in to locate all the centres for the tiny ellipses. Some juggling by eye is necessary when doing this kind of detail at such a small scale.

3. Finally, the remaining details, buttons, lettering, grip details, etc., can be put in. When doing lettering in perspective, you must treat the words in exactly the same way as any other detail, and put in as many construction lines as you need to define each letter. At the very least this means a line top and bottom divided into the number of letters. Remember that when doing capitals the letter 'i' takes up half the room of other letters. When working in lower-case, you also need construction lines for the ascenders and descenders as well as making allowances for the other 'thin' letters.

For clarity it sometimes helps to shade key areas as you go along. This helps your mind to read the drawing as a three-dimensional object in space and establishes one part as being 'in front' of another; and it therefore helps you decide whether the perspective is right.

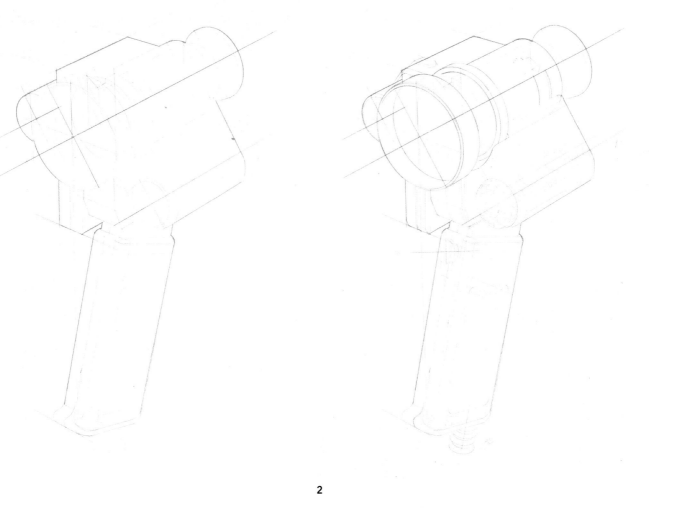

1   2

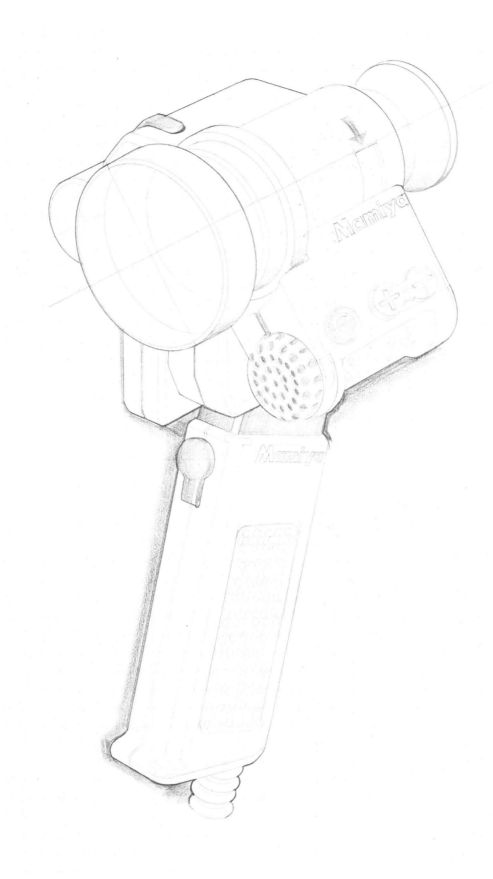

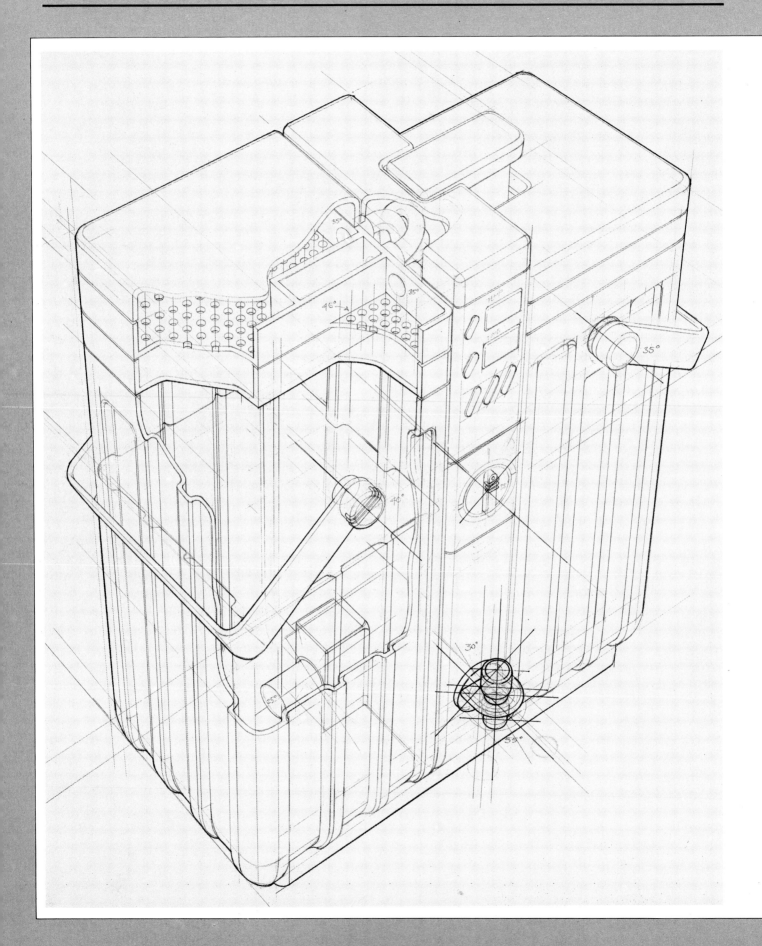

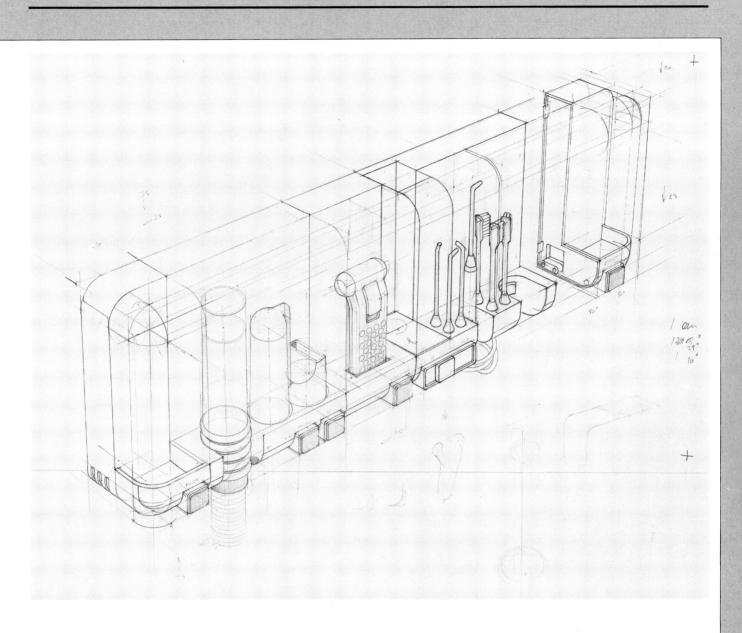

Left: Underlay for beer-making machine
(finished rendering p. 140)
This drawing was done in pencil on layout paper and was set up by eye. Note how the cut-away area is kept to the minimum to reveal as much as possible without incurring too much extra drawing.

Above: Underlay for bathroom unit
(finished rendering p. 99)
This underlay was built up using the cube method from a commercial perspective grid; once the matrix was complete, the grid was discarded, and the details put in by eye. Note how the radii at each end make the product look smaller, and how the angle of each ellipse is marked for future reference.

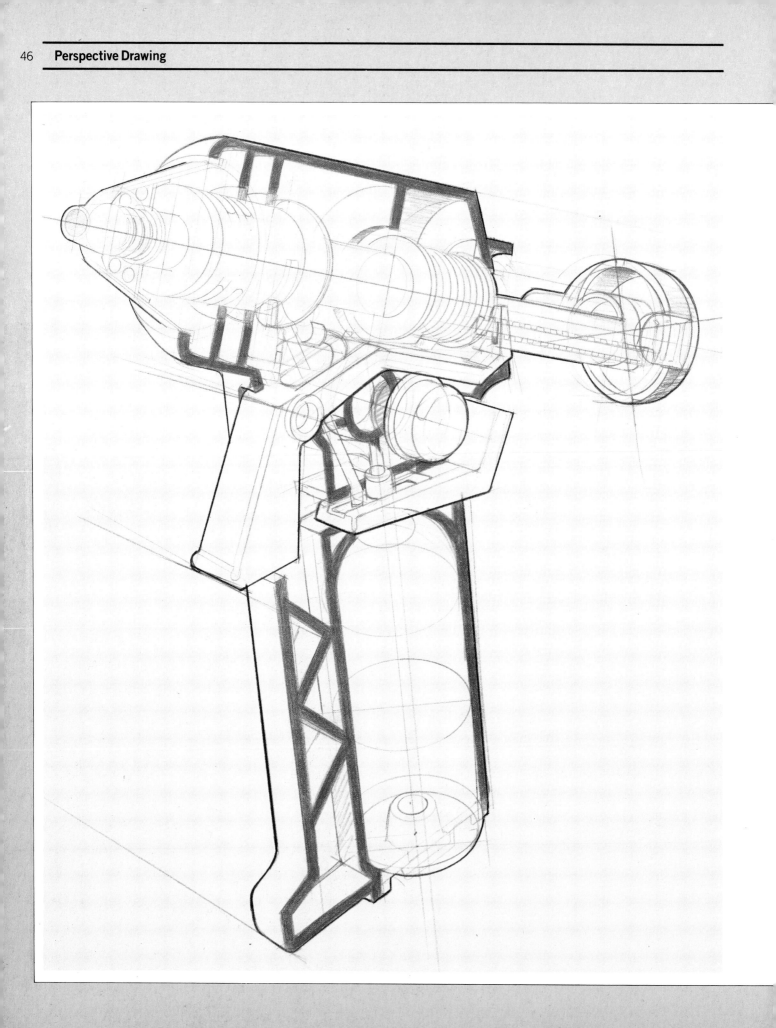

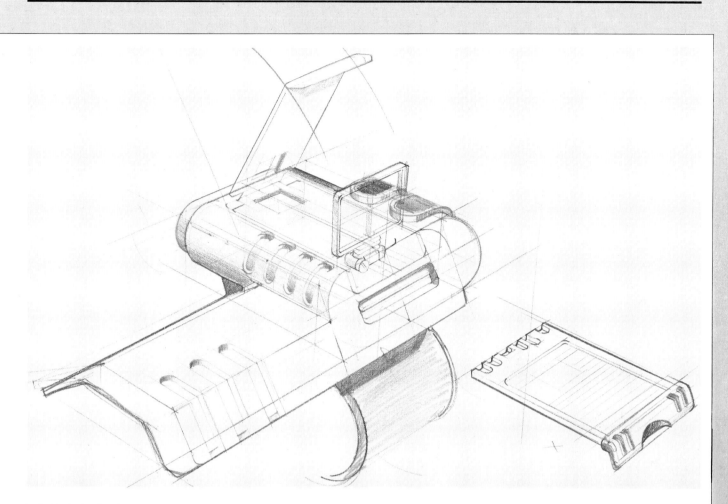

Left: Underlay for the glue-gun

(finished rendering p. 80)

The drawing was built up from the main cylindrical axis using ellipse guides. Note how the major axis of each ellipse also defines the section through the product. The ellipses on the far right are, of course, defining true ellipses on the product and therefore have a different minor axis.

Once the main cylindrical form was established, the handle, trigger, tank, etc. were projected downwards along the split-line and then fleshed out to the rear.

Above: Underlay for 'X-Ray squad' wrist unit

This underlay for a research board was drawn very much off the top of the head. The view was built up by eye using coloured crayons. This is a good technique to use if you find that you build up a lot of lines and then find it hard to distinguish the right one – simply change the colour of the crayon as you work.

Right: Underlay for Sony Walkman

This drawing was originally done for a student demonstration and was traced off a black-and-white photo. If you use this method, make sure that you true up all the lines, and locate the centres and axes of all the ellipses.

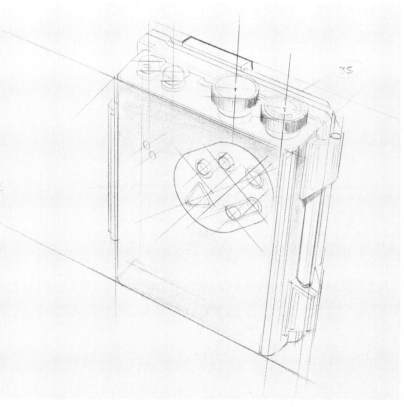

4 Colouring Up

After you have achieved a line drawing of your product design, you are faced with the problem of what to do next. You are probably also considering what colours and finishes to use for the product and working against the clock to meet a deadline – the pressure is on. Before reaching for a marker, however, you need to be in the right frame of mind and you need to plan a great deal in advance.

Attitude

The most usual pitfall to avoid is being 'precious' about a drawing, to a point where you hold back and hesitate because of the risk of making an error. Obviously the nearer a drawing gets to being finished the more precious you will become. The first thing therefore is to be *bold*. Rendering is, in any case, a gross (but effective) simplification of reality; being bold about colour choice, tone, highlights, etc. actually makes it easier. It's like looking at the product through half-closed eyes, which eliminates all but essential details. If you are in doubt about anything, for example, how dark to make a shadow, or how light to make a highlight, then go several degrees darker or lighter than you judge is right.

If you are a beginner, approach the rendering in the knowledge that it will probably take two or three goes to build up your confidence to a point when you can attack the final version. For this reason, never work directly on your master underlay and, if possible, practise on photocopies or dyeline prints so that you don't have to constantly redraw it. Even when you are just sketching through ideas, try and maintain a fluid approach and never be afraid of consigning an effort to the bin. If you don't try, and trying means experimenting a lot, then it will take you a lot longer to improve.

The second attitude to develop is a strong sense of graphic balance. Every drawing you do is a complete graphic image which, like a painting, is subject to simple rules of composition. Obviously, the view you choose is very important but consider also how it is placed on the page and whether the colours are working well together. Look at the size of the image in relation to the page, and whether the general presentation of the drawing looks good. There usually comes a point where the overall quality of the rendering begins to take over from its original purpose, which is to present a design concept to your client. Don't worry about this because, to begin with, it is not an unhealthy attitude and in any case, as you become more confident, your sense of graphic balance improves to a point where it becomes almost automatic.

Probably the most important key to good rendering is observation. The more rendering you do, the more you will look at products around you and begin to understand why they look the way they do. This in turn will improve your ability to determine the disposition of tone, reflection and shadow in your drawing, and the better you get at this the more you will appreciate just how important keeping your eyes open really is. You will find yourself looking at objects in a different way, trying to work out why there is a reflection here or a highlight there. You will begin to understand why we perceive a colour as we do, and how it changes depending on its surroundings and the lighting. In absent moments you will probably find yourself looking carefully at complex reflections trying to work out exactly what is being reflected where and why.

The greatest single effect of your improved understanding of why things look the way they do will manifest itself in your design work, because you will begin to appreciate form and how shapes relate to each other; you will be able to see where formal decisions have failed in existing products and why a successful design works so well. Your visual vocabulary, or 'visual experience', will be greatly expanded, which will allow you to command more design options and so improve your ability as a creative designer.

The final point to bear in mind is economy. Remember that as a designer you are trying to give an impression of reality, rather than portraying reality itself. You want to put across an idea or design to your client, not impress him with the quality of your draughtsmanship. There is therefore no need to be absolutely faithful to rigid lighting conditions and accurate reflections; concentrate instead on using only those elements which help describe the finish or form you are trying to illustrate. The conventions which follow below may be clichés in the design world but that is of no importance to the client. Provided the client interprets the drawing in the way you intend, then you have achieved your objective.

Planning

Too many people believe that it is skill with this or that media that makes a designer good at rendering. Of course, this is important, but not as important as really understanding how reflections, colours, highlights and so on work. Designers who understand why we see things the way we do can turn their hand to any media, because they know what to draw. After that they will need constant practice with a new medium to perfect their rendering; as with perspective, practice really does make perfect. So there is no need to worry about your intended media too much at this stage.

Getting back to the blank sheet of paper, or uncoloured underlay – where do you begin? There are two basic approaches to be considered: the first, the more traditional way, concentrates on imagining the product illuminated by a single light source usually behind and to one side of the viewer, i.e. over the left- or right-hand shoulder; the second is to consider the product in terms of reflections.

The first method is indispensable for determining shadows and highlights but it is misleading, in my view, to think of the light as 'coming' from anywhere in particular. This is because the actual colour we perceive in a product, be it a car or vacuum cleaner, is a function of four variables:

1. The intrinsic colour of the surface.
2. Its reflectivity and finish.
3. The colour and tone of its surroundings.
4. The location and intensity of light sources.

Of these four variables the first is a design decision while the second and third are more interdependent than the fourth. We shall come back to the exact light source later in the chapter but for the moment we will concentrate on reflected light. We must look at the line drawing and decide what is being reflected and where, and how best to portray it.

To do this we shall first consider some basic shapes and observe how their form is described with reflections and how matt (non-reflective) surfaces are subject to the same conditions. We shall also illustrate and explain some of the conventions that you can use in your own rendering. We shall consider each shape at its most reflective: i.e. mirror, or chrome, and then observe how the reflected image is identical in form, but not colour, to that seen in a gloss-plastic version; and we shall look at a totally matt version and see how it responds to the same reflected images.

Cube

The chrome cube has no colour; the colours we see in it are entirely a function of its surroundings. Each surface is effectively a mirror which reflects in full colour and without distortion objects, landscape and light sources around it, but the 'picture' we read in each surface represents only a limited part of the surrounding environment. The easiest way to work out what is happening on these surfaces is to make a cardboard cube and laminate some mirror-finished polyester film, like Mylar, to each face, or simply hold a small mirror as if it were the side of the cube. Try placing it on a sheet of gridded paper, and move coloured shapes up to the surface, watching all the time what happens to the reflections. Try and cast a shadow onto the mirrored surface and you will find it very difficult. Observe how anything at right angles to the surface appears to pass right through.

You can see that if you wanted to draw a chrome cube you would first have to imagine it in its surroundings, and probably actually draw some of those surroundings. (To find out how to do this, refer to p. 57.)

If we now imagine a highly polished, gloss-plastic cube, it too reflects in exactly the same way as the chrome cube. The same images we saw in the sides of the chrome cube can

(To find out how to do this, refer to p. 57.)

be seen just as crisply delineated in the plastic version. They are also sharp, pictorial reflections, but, unlike the chrome cube which reflects in full colour, they are made up out of the tonal range of the cube's intrinsic colour. Test this by finding something with a flat, polished and coloured plastic surface (a piece of coloured acrylic is ideal) to compare with the mirrored cube above. Take a red, yellow and blue coloured pencil and try holding them at right angles to the surface of the mirror – the reflected image is perfect in every detail including colour. Next, try them against the polished plastic – the shape and crispness of the reflection is the same but most of their intrinsic colour is lost. Exactly how much of this colour is reflected depends on the colour of the base material, but for rendering purposes you can ignore the subtleties of reflected colour and use a tone of the base colour. There is one small exception to this guideline and that is gloss black, which reflects a lot of colour. This may be one occasion therefore when you need to go further than simply using the tonal range of the base colour. The more experienced and

confident you get, the more you will begin to exploit the potential of reflected colour; but for the moment it is best to ignore it.

Remember also that, with super-glossy, coloured surfaces which have been lacquered, the more acute your angle of view to the surface the more reflected colour you will see, and the closer it comes to a right-angle the more intrinsic colour you see; at low angles of incidence the glossy lacquer is very reflective and you see the surroundings, but when viewed directly from above you see right through the lacquer to the underlying colour. It is important to understand this when rendering glass and other transparent, or semi-transparent, materials, as they behave in exactly the same way.

So, to draw a glossy plastic cube you need to think of it in exactly the same way as the chrome cube, which means considering each surface in terms of its reflections. At the other end of the scale you should study a matt-black cube to see how it might look in similar circumstances. Before you do, however, it is important to understand why the matt surface looks matt and why the

Cube

1. This is a schematic representation of what you would see on the flat surfaces of a chrome cube. In each surface there is a specific area of the surroundings unaffected by distortion. In the top you will see an area of sky, or ceiling, and in the side you will see an area of the base immediately in front of the cube.

sky tones

ground tones

1

polished surface is reflective. The polished surface (and the mirror is the perfect example of this) is absolutely flat. If you look at it under a microscope, you will see that the absence of surface aberration allows light to reflect off it in a coherent and predictable way. The matt surface, however, when seen under the microscope, is finely textured and pitted, which scatters the reflected image in all directions. This makes it impossible to read the reflected images because they become blurred and indistinct. Only the brightest of images, such as light sources, will be seen.

If you repeat the experiment with the coloured pencils against a flat matt surface, it is very difficult to see any reflection. Next, position a light source so that you can see its reflection and watch how the image which is sharp and crisp on the glossy surface is blurred and diffused on the matt one. This effect is especially important when drawing matt finishes because highlights (that is concentrated areas of reflected light) are ill-defined and soft compared to their glossy counterpart.

To sum up then, you should consider each surface of the cube separately, whatever its finish, in terms of what is being reflected. I usually treat the top, upward-facing surface as the lightest because it reflects the sky, if outside, and the ceiling complete with lights, if inside. Both the remaining faces will be darker and one will be darker still. Usually the larger of the two gets the lighter treatment but this can depend on which side of the product needs most emphasis. You can apply this method to any rectangular product even if it is multi-faceted – just remember that every plane which faces in the same direction will reflect approximately the same information and will, therefore, be similar in its tonal value.

Cube (cont.)

2. In this chrome cube, shot in a photographer's studio, the top surface, apart from the black hole, is reflecting the 'sky' (actually a large light) above, and the two sides are reflecting their respective parts of the gridded rectangle and the beige paper on which it sits. Note how, on the near right-hand corner, the reflection of the rectangle is 'pulled', or distorted, into the edge giving it an almost liquid appearance.

3. The view of this chrome cube is the same as that used later to construct reflections (see pp. 57-8), so that you can see how these were built up. It is impossible to give a good impression of a chrome cube without drawing at least some of the surroundings. In other words, if you were to take away the block on which the cube sits, the viewer would make very little sense of the reflections. The top is reflecting sky, and the hole through the middle is reflecting ground at the top and sky at the bottom; in between is the horizon. The reflections are crisp and sharp, and the corners (which, like those of a radiussed cube, focus and compress the surroundings) are strongly contrasted.

4. With gloss plastic it is not really necessary to go to all the bother of constructing the reflections, as a good impression can be obtained from a more general approach. (However, if you had placed the cube on the same block as before, then you would see exactly the same reflections but in tones of blue). Each face is treated separately: the top is reflecting sky, or ceiling, and has a vertical window reflection running across it. The nearside face is considerably darker and is graded slightly so that it is lighter at the bottom. The face with the hole through it is slightly darker still and is also graded towards the bottom; this is to give the most contrast at the top edges where the highlight runs; the

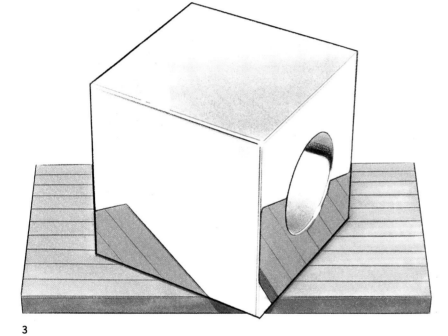

3

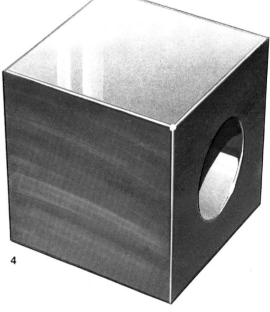

4

highlight itself is backed either side with a dark crayon to make it appear brighter. The hole is exactly as in the chrome cube but with a shadow across it.

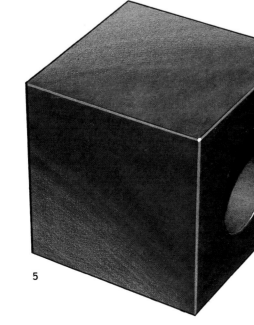

5

5. With the matt-black cube all the reflections are lost leaving only a gradual tone across each face. The top face is lighter still and slightly graded from front to back. The hole now has a fuzzy instead of a sharp highlight although the shadow across it is still crisp.

Cylinder

Having seen how a completely flat surface reflects, we must now look at what happens when a surface is curved in a single plane. This is best explained by looking at a simple geometric cylinder in the same context as the cubes.

Horizontal cylinder

1. This is a schematic representation of the 'desert cliché' – what you would see in the side of a chrome cylinder with the sky at the top, the ground towards the bottom and the horizon in-between.

2. This cylinder was photographed in a studio but illustrates the desert cliché well. We were unable to simulate the sky, so, as with the cube, we had to use a large light with a piece of blue graded paper tacked onto it. Note how the flat end of the cylinder reflects an undistorted image of the surroundings.

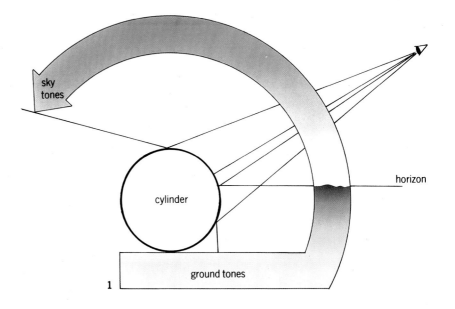

sky tones

cylinder

horizon

ground tones

1

2

We have seen how chrome, because it is the most reflective of materials, can give us all the information we need to understand other finishes. So let us consider first what we might see reflected in the surface of a chrome cylinder. This is best understood by imagining a chrome cylinder laid horizontally in the artificial and uncomplicated environment of a desert, with you, the viewer, looking down at it. What you see in the surface of the cylinder is a distorted view of the surroundings. At the top of the curve you see the sky, in the middle you see the horizon, and as your eye scans the lower part of the curve you see the desert floor being reflected. As your eyepoint moves up and down, so the horizon moves up and down as well. In very reflective surfaces, like chrome, there tends to be great contrast between lights and darks, so imagine that it is dawn or dusk in the desert, with the sun low on the horizon. This throws the horizon into dark shadow and, with the sky above it at its lightest, produces a very high contrast. The colours in the reflection, as in the cube, are exact and true to the real image. I call this reflection the 'desert cliché', and it can be applied to chrome in almost any situation. You can see it in any glossy curved surface, especially in the sides of cars where glass and polished paintwork reflect strongly. Often it is not the true horizon (in a perspective sense) that is reflected but the roofline of adjacent buildings. It is perhaps best to think of it as a positive and high contrast change between light and dark.

Of course, many products are for interior use, so there is unlikely to be a true horizon to be reflected. However, the effect is still easy to see, often as a table edge, or simply the border between light and dark. Most commonly it is a window being reflected and because the window is usually the brightest part of the room the wall in which it is set is the darkest. This, as in the desert, provides the high contrast and 'edge' that makes reflective surfaces look reflective.

Turning the cylinder through 90 degrees produces a very different reflection, when viewed from above. The horizon will not be visible until you drop your viewpoint to horizon level; instead, the surface reflects a distorted view of the desert floor which appears to wrap around the cylinder. This is best observed by taking a small section of chromed tube and placing it, end-on, on a piece of gridded paper. With tubes of a large diameter the reflections can become quite complicated (see photo on p. 54), but with thin tubes of small diameter the reflections typically resolve into a series of dark and light verticals. In tubular chrome furniture, for example, it is often best to suggest the

reflective finish in the horizontal elements and at the changes in direction and leave the verticals almost untouched. Alternatively, it is sometimes possible to get away with the desert cliché on verticals with no loss of credibility to the drawing.

The gloss-plastic cylinder, of course, whether horizontal or vertical behaves in exactly the same way as the chrome version but, like the cube, reflects in tones of its intrinsic colour rather than in true colours. The horizon will probably be the darkest area, and the sky will be reflected as a light tone at the horizon, getting slightly darker as it wraps around. Below the horizon the desert floor will be a dark tone becoming lighter as it wraps around away from us.

The matt cylinder reflects exactly the same environment but, as with the cube, the information is blurred and indistinct. The light source, in this case the area above the horizon, will be seen as a hazy highlight running along the cylinder. The graduation of tone around the surface will be much more subtle and appear as a consistent change of value as it wraps around.

3

4

5

3. In this rendered visual of the chrome cylinder you can clearly see the desert effect. The flat end did not look quite right with just a flat sandy colour being reflected (it is difficult for the viewer to make sense of the reflected image without seeing the real one) so, since chrome appears predominantly black and white, this was left white with some blue to provide contrast to the highlight around its circumference.

4. In the gloss-plastic version the same desert effect, with a sharp horizon, can be seen. Note how, like the chrome version, the horizon is 'pulled' (distorted) into the little radius where the highlight runs around the circumference. In this example a darker tone in the flat end is more appropriate to give contrast to the highlight, and this is slightly graded from top to bottom.

5. The matt-black cylinder shows the horizon/sky reflection resolved into a smooth tonal transition.

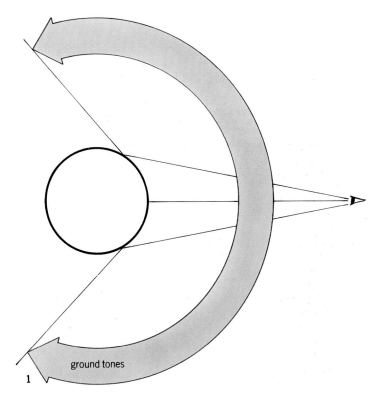

1

ground tones

3

4

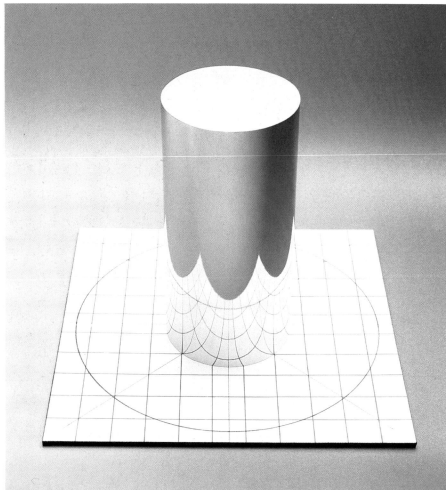

2

Vertical cylinder

1. Maintaining the high viewpoint on the cylinder but turning it through 90 degrees produces a very different effect. All you can see in the cylinder is a distorted view of the desert floor. If you lowered your viewpoint so that it was perpendicular to the vertical face, you would be able to see almost all the horizon, with yourself in the middle foreground.

2. The photo shows how, with a fairly high viewpoint, the surface of the cylinder reflects the ground around it. You can see how even the two back corners of the square are just visible. Note, too, how the red squares and the circle drawn on the base, are reflected.

3. The rendering of the chrome cylinder shows how it is possible to use the desert cliché without losing credibility, and without the need to draw in the base. For maximum realism it should have been drawn, like the cube, on a surface which could have been reflected. In practice, though, this is extremely difficult to do and you are as likely to get it wrong as right. As a compromise I have, in the past, used the distorted corner (as in the vacuum flask on p. 85) to suggest a base surface.

4 and 5. As with the other examples, the gloss-plastic cylinder is exactly the same as the chrome but in tones of blue, and in the matt-black version the highlight is resolved into a hazy, but smooth, tonal transition.

Sphere

The sphere combines the effect of the cylinder in a horizontal and vertical position, producing a reflected image of the surroundings which is distorted, like a fish-eye lens. The reflections appear most distorted near the edge of the sphere and least distorted on a line drawn from your eyepoint through to the centre. As with the previous examples, it is easiest to imagine a chrome finish in the desert and work out in side and top elevation exactly what is being reflected. You will see the high-contrast horizon relatively undistorted in the middle but being 'pulled' and 'stretched' as it nears the edge and your view is more tangential to the surface. In the immediate foreground you will see the desert floor and you, the viewer, and right at the bottom a reflection of the sphere's own cast shadow, while above the horizon is a complete vista of the surrounding sky. As with the other examples, the gloss-plastic sphere reflects the same images but in tones of its true colour and the matt-black

version resolves all the reflections into a smooth tonal transition.

Reflections in any compound curved surface are difficult to predict but as your experience builds up you will become more and more adept at it. It is very useful to keep a selection of reflective shapes which you can refer to; for example, a polished billiard ball or a silver spoon (which gives both convex and concave reflections). I also keep a couple of mail-order catalogues because there is endless visual reference in there for all kinds of shapes and finishes.

Sphere

1. The reflections in a sphere are a combination of those seen in the vertical and horizontal cylinders. This photograph of a sphere, shot in a studio, was specially set up in order to simplify the reflections. (Since the sphere reflects like a large fish-eye lens, the photo would otherwise have shown the entire studio complete with the photographer and his camera.) Note how the red squares are distorted as they wrap around the surface.

5

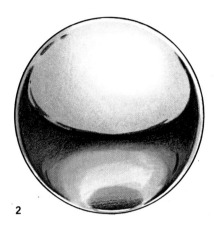

2

3

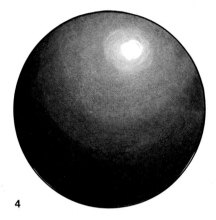

4

Sphere (cont.)
2. The chrome sphere shows the horizon being distorted as it nears the edge, with the desert floor being reflected in the lower hemisphere. In this case the sun (highlight) is higher in the sky and can be clearly seen.

3 and 4. The gloss-plastic sphere is, of course, identical but in tones of blue, while the matt-black one is continuously toned from the highlight outwards.

Radiussed Edges

Armed with these three basic examples of how reflections work we can begin to combine them into more product-like shapes. Perhaps the most typical of modernist forms is the flat-sided geometrical shape with radiussed edges. In the previous chapter we constructed a cube with radiussed edges and I referred to the usefulness of this constructional technique in determining the disposition of reflections.

First, consider the radiussed cube in side elevation, breaking it down into flat, cylindrical and spherical surfaces. In the flat side of the cube you will see, like a mirror, an undistorted image of another section of the surroundings (probably the ceiling or sky). Between these two areas, the radius is reflecting a total picture of the remaining environment. If you could walk across the surface, you would see every detail of those surroundings; the radius has the effect of focusing this large amount of information into a very small area. This is why radiussed edges often appear very bright, as highlights.

Note that in the side elevation the rate of change from flat surface to radius is gradual at first so that the horizon (or highlight) does not occur at the line where the radius starts, but some way up its surface. Note also that if this were a matt-finished cube we would read one tone on the top surface and one (probably darker) in the side. Between the two there would be a gentle and consistent

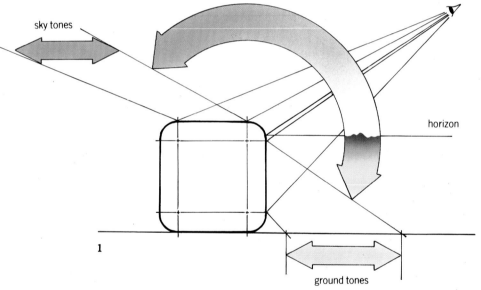

sky tones

horizon

1

ground tones

graduation to a bright but hazy highlight.

Apart from the desert/horizon convention there are some other useful and easily controllable reflections which can be used to advantage.

Radiussed cube

1. This schematic diagram illustrates what you might see if you looked down upon a radiussed cube. In the front face there would be a reflection of a limited part of the surface on which the cube is sitting; this would be distortion free. In the flat area of the top face you would see an undistorted image of the sky, or ceiling. In the radius between these two flat planes you would see everything else. This is why you tend to get highlights on

radiussed edges; for example, if the sun was in the sky behind you, you would see it being reflected in the radius even if you moved your viewpoint lower down or higher up. For the same reason the radius has a focusing effect, picking up many individual lights (spotlights on a track for example), and resolving them into a band of light that is brighter than the rest of the cube.

A common mistake among students rendering radiussed edges for the first time is to draw the highlight as broad as the radius. Because the curvature of the radius from the vertical is gradual, the horizon normally lies slightly above the imaginary line which defines the end of the flat plane and the beginning of the radius.

2

2 (Above). Tonally, this gloss-plastic radiussed cube is the same as the one on page 51 with the top reflecting sky, etc. The radii are treated as three broad highlights; note how they are 'pulled' in that characteristic liquid look as they reach the back corners. Giving all three highlights equal brightness is the easiest way to render the radii, but they could have been treated differently, with the vertical probably being a similar tone to its two adjacent sides (it is, after all, a vertical cylinder reflecting the ground on which it sits) and the highlight sweeping around the top only.

Reflections in a Flat Surface

As we have seen, a flat reflective surface behaves exactly as a mirror. That is, it reflects a true, undistorted image of a fixed part of its environment. It is easiest to understand this everyday effect by positioning a pencil at 90 degrees to a mirror and looking at the reflection. It appears as if the pencil passes straight through the mirror surface, and the apparent length of the reflected pencil is the same as the original pencil (in perspective, of course, because of the effects of diminishing distance). This effect is the most predictable of all reflections and can be used to construct more complicated reflections on those occasions when absolute accuracy is important. The adjacent illustrations show how to construct simple reflections in the side of a cube.

If, however, you had chosen to sit the cube on a small circle (which, in perspective, appears as an ellipse), then, after constructing a circumscribing square around the ellipse, you could use the same basic technique as that illustrated to construct the reflection of the circumscribing square. Once this is done it is very simple to lay in the ellipse within it.

Below: Constructing reflections in a flat surface

1. To get the feel of this single and obvious reflection, construct a perspective cube (as described in the previous chapter) sitting on a block. We need to work out where, in each face, the reflection of the block will appear.

2. The easiest way to do this is to treat each vertical face completely separately and, whichever one you choose, imagine it as part of a much larger mirror and extend its base line accordingly. Ignore the rest of the cube for the moment and concentrate exclusively on this 'mirror face'. Mark off a series of lines (shown in red) along the base, that pass through the mirror face (shown in blue) and are at right angles to it in perspective. This means they will share the same vanishing point as adjacent faces of the cube. Where a line crosses a detail on the base, in this case the edge, a point (X) can be marked off; you

can then estimate the position of the reflection of this point (X^1), as it will lie on the same construction line and at an equal distance in perspective on the other side of the base line. It is as if this point has been rotated through 180 degrees to give a symmetrical point about the base line of the mirror. Remember that where a line actually meets the mirror face (A and B) its reflection starts at this point.

It does not matter at this stage where you draw the red/blue lines because you quickly learn to recognize the important points to be reflected and put the lines through these key points. In this example, you do not really need all the red/blue lines, only the one which runs through point Y, which helps you to locate Y^1. Joining Y^1 to A and B would construct the reflection. If, however, line YB was wavy or there was a complex pattern on the base, you would need all of the red/blue lines. Note that the thickness of the base is also reflected in the mirror.

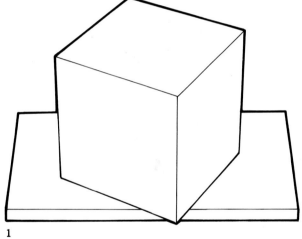

1

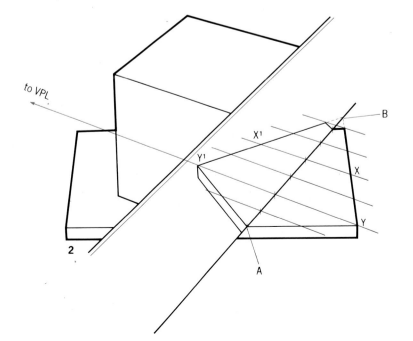

2

Constructing reflections (cont.)

3. Finally it only remains to reduce the mirror face back to its real size, rubbing out all extraneous lines in the process and just leaving the relevant part of the reflection.

4. Repeat the process for the other face of the cube.

5. Reduce the mirror face back to its original size, again eliminating unwanted construction lines.

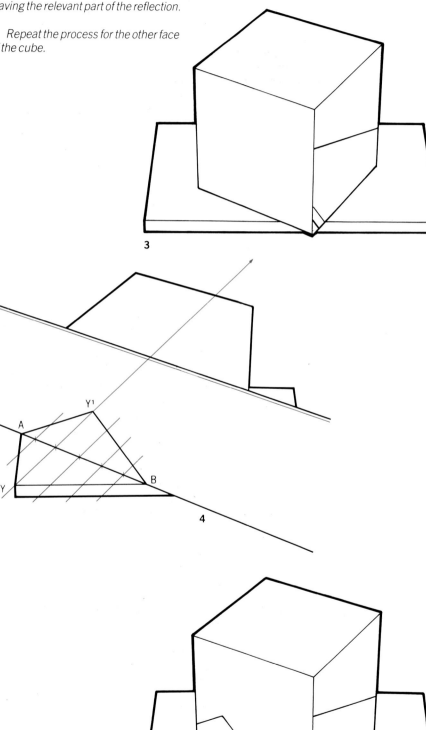

Remember that, provided the mirror face and surface being reflected are at right angles, then the reflected image shares the same vanishing points as the real image. Once the mirror face is rotated from the vertical, then the reflected image is also rotated in the same direction and all the construction lines should bend accordingly.

This basic technique can be easily used in two other clichés. The first is placing the product on a surface. Depending on the type of product, choose a surface that is easy to reflect. The most commonly used are tiled walls and floors but nearly any abstract pattern can be used to good effect. The second is using a part of the product itself. Rather than imagining the environment, or actually drawing a selected part of it, you may be able to use an element of the product itself to effect a bold reflection. For example control knobs, handles, folding covers or any moving parts may be positioned so that we see their reflection in adjoining surfaces. This allows us to choose the best possible reflection and to construct it as easily as possible.

The Window Cliché

The window cliché is one of the most commonly used conventions to make surfaces look reflective. Cartoonists use it extensively, as do photographers when setting up product shots (although, of course, they simulate it by using a 'fish-fryer' light).

I use it most in horizontal surfaces. Nearly all windows are set in vertical walls and usually the wall in which the window is set is the darkest in the room; this provides the high contrast necessary for making things look glossy. Any horizontal surface, such as a table, or the top of our cube, will reflect the window as a high contrast *vertical* band (any line at right angles to the mirror surface will appear to pass through it). If the surface is slightly angled then the reflection will also be angled. You can therefore use the window reflection to indicate changing planes as, for example, in a record turntable top.

Below: Reflecting a part of the product

1. To make a surface look reflective it is often possible to position a part of the product so that its reflection can be seen in another part. In this example the open lid is reflected in the top of the cube and the shape on the side is reflected in that side. If the shape is itself reflective, then this will have a reflection of the cube in its side. As a general rule (and contrary to expectation) the reflected image is slightly darker than the true image.

2. It can be very effective to place the product on a surface or in an environment that reflects it – although this says more about the surface's finish than the product's.

Bottom: How highlights work

Highlights, or 'chings', are the reflections of bright light sources – the sun if outside, and lamps if inside. The areas in shadow are those areas diametrically opposite the light source.

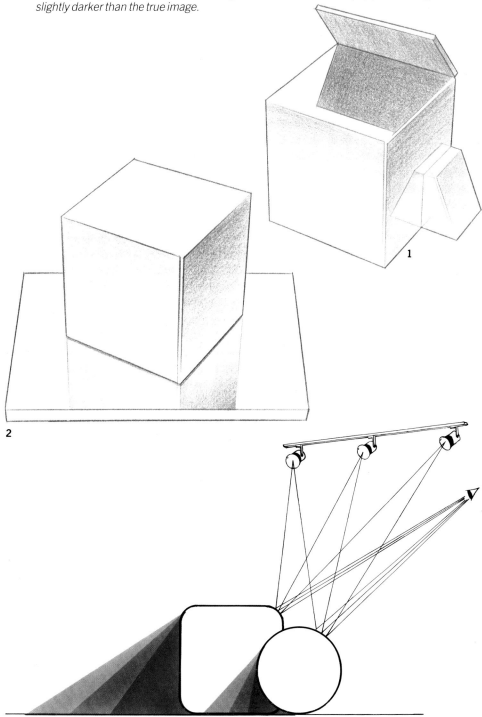

Highlights and Shadows

It is appropriate at this stage to return to the second basic approach outlined at the beginning of the chapter and imagine the product illuminated by a single light source. I usually position this almost directly over my head and to the left or right depending on which side of the product has most interest.

You can of course put the light source anywhere you choose, to obtain more dramatic effects, but once you do this you increase the shadowed part of the drawing considerably. Backlighting, for example, is difficult to do and throws the bulk of the product into shadow making it difficult to render (no highlights) and more difficult for the viewer to understand. Remember also that the focusing (or catch-all) effect of the radii that 'fall away', or point away, from the viewer are less pronounced, so that even the brightest, broadest light source behind will yield only a pencil-thin highlight on a falling-away edge. Highlights, sometimes called 'chings', are the reflections of the light source itself. In the case of an exterior view this would be the sun, and with an interior view will probably be a single spotlight or track of spotlights.

When putting in highlights, do not think of the light as coming from anywhere, but rather think of it in terms of where on the product the light source will be reflected. With the light source, whether sun or spotlight, almost directly over your head and slightly to one side, all the near edges and curves will catch the highlight, and their focusing effect will be at its most pronounced.

It is also, of course, the light source which determines the disposition of shadows, so it is convenient to consider the two effects together. With the light source above and behind the viewer, the shadows are thrown behind and below the product, so that they have an absolutely minimal effect on the ground or base. Where they do become crucial is when the product has a part (or parts) throwing a shadow across itself. In this case it is nearly always essential to put in the shadow and it is usually not difficult to estimate its position.

Remember that the colour of the shadow appears as a dark tone of the material on which the shadow is cast. This means that, if a shadow is thrown across two different materials, you should change the colour you use for each part as it passes across the two materials. Remember also that there is virtually no shadow on chrome.

5 Marker Rendering

Of all the materials available to the designer the marker is probably the most widely used. It has spawned a wide range of techniques because it is fast, easy-to-use, compatible with other materials and gives strong punchy results. It has also done much to break down the barrier between working sketches and finished drawings. This, in turn, means that designers are less likely to finish the design work and then sit back and ponder on how best to do the finished visuals, because they will have been working with colour from a much earlier stage in the design process.

Although designers refer to 'marker rendering' as a single technique, most visuals of this type rely heavily on other media as well, such as coloured crayons, pastels and paint. Each marker only contains a single colour which, when applied, leaves a defined edge and therefore makes the rendering of soft, gradual toning very difficult to achieve. So, for soft organic shapes it is perhaps less suitable than pastels or coloured pencils.

The marker itself is an unforgiving medium and cannot be erased if wrongly applied. Paradoxically, it is also a medium that demands a *bold* approach for best results. Because it is so fast and so immediate, it is easier to avoid the 'precious' approach to a drawing. Too many students use the marker indecisively and become over-concerned at slightly blurred edges where, perhaps, they have strayed over the line. This approach produces dull, static drawings instead of fluid, loose visuals which always look more effective. Many beginners will follow the same course as I did, starting off too tightly and then gradually loosening off as you build up experience; try and short-circuit this process by constantly practising a more fluid approach.

Marker rendering, more than other techniques, is best at giving an impression, rather than a true picture, of reality. This impression of reality is similar to that obtained by screwing up your eyes when looking at an object. You are trying to distil the essence of the object by simplifying the tonal values and details, in order to create an impression (or series of clues) with which to reconstruct an image close to reality. Provided enough information is supplied, the mind can draw on visual experience to fill in the gaps. The really good renderers are the designers who are as economical as possible with their marks on the paper, but still manage to create an informative visual.

To help you achieve this economy it is a good idea to *exaggerate* what you would see through half-closed eyes; in other words, simplify the image, and then render it with more contrast and bolder strokes than you judge would be there in reality. One pitfall for the beginner to avoid is overworking the drawing. Learn to be economical and resist the temptation to cover every square inch with the marker. Often it is more a question of where *not* to put colour, rather than a question of where to put it. In fact, the surface of most marker papers will not stand too much overworking and will quickly reach saturation point; scrubbing of the marker on the paper will also quickly break up the surface.

Try to avoid a rigid, stilted approach to the drawing which is usually the result of lack of confidence and a hesitant hand. The marker really demands drawing from the elbow and not the wrist. You have to keep it moving just to avoid 'puddling', or flooding the paper, so, as a medium, it doesn't suffer fools gladly. There is only one solution, and that is constant practice. The more you practise the better you will become, and the better you become the more confidence you will have.

Basic Techniques

Using the tip. Use the chisel tip to provide a range of line thicknesses. When working the marker, be sure not to let the tip dwell on the paper where it will form a blodge.

Infilling flat areas. This is the most basic use of the marker and the starting point for most beginners. It is very important that you begin by mastering this technique because, once you have acquired the necessary control to do it, everything else is simple. Try to work freehand (without rulers and masks) as often as possible to help build up confidence.

The key to obtaining a flat finish is the maintaining of a 'wet front'. This means that you must work fast so that the front edge of ink does not have time to dry out completely. To do this you will have to backtrack constantly over the front edge to keep it wet and on the move. Before laying down a flat area of colour, try and plan out a route for the marker that will allow you to maintain this wet front. If this is not possible try and leave the wet front at a joint-line or change of direction where it will not be noticed.

Marker streaking. The experienced marker-user often uses the 'streaking' quality of the marker to good effect. It can help describe a form, suggest a reflection, relieve the visual monotony of a flat surface or simply give the drawing some direction. Many designers maintain stocks of run-down markers, with which they can obtain a wider range of streaking effects than is possible with full markers. I recommend that beginners master the application of flat colour before going on to exploit the potential of streaking. One of the first areas where the novice will want to exploit streaking effects is in the rendering of round, tubular shapes which are difficult to do in any other way.

Masking. Masking is done with either low-tack masking tape or masking film but be sure to test your paper first to ensure that the surface is undisturbed by the tape or film. It is a very useful technique where a bold, streaky finish is required to finish at right angles to a line, but without the hardening of edge that normally results if the marker is gradually decelerated to a halt. The masking allows the flow of the streaks to continue uninterrupted.

It is much safer to work across the tape or film than along it as this helps prevent bleeding under the mask. When working with giant markers it is usually essential to mask.

Below: *To fill a square like this with flat colour is not as easy as it looks, because you must work fast and with confidence. The photograph shows six intermediate stages starting at the top left: turn the square so that the left edge is horizontal and lay a strip of colour up to the line, turn the square back and lay another strip along the top line. Take the marker back into the left corner and work it outwards in diagonal passes ensuring that you go over the two lines often enough to keep them wet. Put in a strip of colour along the bottom edge (turning the page if you are not too confident of staying within the line) and infill from the bottom-left corner keeping the leading edge wet. Finally, work this wet edge across to the right-hand line.*

Right: *In this illustration of marker streaking a series of Warm Grey markers have been used to build up a cylindrical shape.*

Right: *When using masking tape do not expect a super-crisp edge because the marker will nearly always bleed very slightly under the tape, especially if it is allowed to dwell near its edge. The technique works best if the marker is used boldly across the masked area, as here, rather than carefully as when infilling colour.*

Working to a line. On all but the longest lines practise working freehand as often as possible. Try to work to the line in one continuous sweep rather than a series of short jerks. Above all, don't worry if you go over the line or miss it completely in places. A line drawn afterwards with a coloured pencil will tidy it up effectively and the eye will read this line in preference to the jagged edge.

When working with a ruler use the marker quickly to prevent ink being drawn under the ruler. Be especially careful when using new, well-filled markers with soft nibs where there is a risk of bleeding. On these occasions work with the ruler bevel-side down and/or positioned on a part of the drawing that is less important. Most plastic rulers are attacked by the solvents in markers and will eventually deteriorate; this is not usually the case with the PVC straight-edge described on p. 23. Whatever you use, be sure to clean it frequently with a tissue dampened with solvent. This is particularly important when changing from one colour to another, as there is a risk of traces of the first colour appearing in the second.

Toning and blending. It is quite possible to blend adjacent areas of colour together to produce a more gradual change than is normally obtainable from marker inks. This is only possible on papers with a low absorption rate, such as Vellum, Crystalline and thin layout papers, and is best done before the marker application has completely dried out. Depending on the size of the area being blended, work with a cotton bud, tissue or lint pad soaked in solvent and, if necessary, ink taken from the end of the marker. This is particularly useful for blurring the crisp edge of the marker where it meets the white of the paper. Another technique for colour blending is to slip a sheet of acetate under the paper. Because the acetate is totally impervious, the ink will not be absorbed by lower layers of paper and will tend to form puddles. The whole area being worked can then be kept wet more easily, allowing the colours to be blended.

Overcoating. Depending on the rate of absorption of your chosen paper, one marker is capable of producing two, and sometimes three, tones of its colour. This is the normal and most obvious way of achieving different colour values. To obtain the greatest difference in value, be sure to wait until the first application is absolutely dry before going over it again. With most papers the third overcoat produces very little change and subsequent coats will saturate the paper. This causes puddling which can dry to a

sticky finish. A light dusting with talc will eliminate the stickiness and also 'knock back' the colour to allow a further coat. On some papers (particularly Vellum) further tonal values can be achieved by working on the reverse side of the paper; this is not possible on marker papers which have been treated to prevent bleed-through.

As a general rule do not expect to achieve sufficient tonal range from a single marker; a drawing done with a single marker will lack contrast between lights and darks and therefore tend to look rather flat. Faced with this problem many students reach for the greys to produce darker tones. Working with greys to produce a wider tonal range results in muddy colours and gives a slightly grubby appearance to the drawing. It is far better to assemble a palette of colours that work well together, either on their own, or when overlaid. For example, when working with a bright yellow marker as your base colour use a yellow ochre for the mid-tones and a light brown for the darker areas. Alternatively, if you want a warmer, more vibrant appearance, use a yellowy orange and pure orange respectively.

Look back at the gloss-plastic cube on p. 51 (No. 4) as this shows how to break up the three tonal planes. The top surface is done with pastel and the left face with a Mid Blue; the right-hand face is overlaid with the Mid Blue to darken it further. In addition, Antwerp Blue was used to grade the tone further from the top. Finally, a Prussian Blue was used for the shadow within the hole.

Preparation

First of all, make sure you have all the materials you need to complete the job and that the markers have sufficient life left in them. When you have selected the colour of the product, match it as closely as possible to a marker (often this is not possible and you will have to use two markers to achieve the right colour). Next, find two markers to give you a good tonal range for the darker areas. If you are at all unsure, experiment with them first by doing rough sketches so that you are clear about what goes where. If you intend to use pastels as well, make sure you are clear in your own mind about their colour match.

It is usually a good idea to do a rough first so that you know where the various tones lie and where the reflections will be. Plan everything carefully, at least till you are more experienced.

On the practical side, ensure that the paper you have chosen is suitable (some treated papers don't like masking tape). If you are not working on a bleed-proof pad, be sure to slip a

couple of backing sheets beneath the top one. This is obviously important if you intend to work with solvents as it is easy to ruin a whole pad with bleed-through. If the pad is running out, I recommend switching to a new one because subsequent overworking with crayons will reveal the texture of the card backing-sheet.

Sketch Rendering

The following stage-by-stage examples describe in detail how a particular rendering was produced. They have been divided into two sections, sketch rendering and finished rendering, although the dividing line between the two is fairly arbitrary and depends on the type of work you do. Sketch rendering refers to the drawings you might do as part of the design process and finished rendering refers to the type of drawing you do once the design is finalized, and you only need to present it.

All rendering is about communication – with the client, with your colleagues, and with yourself. As you work, you produce ideas and solutions and the last sparks off the next as you hold a 'visual conversation' with yourself. You must not slow down this process by getting too deeply involved in the details of working up a solution. Later, perhaps, you will go back through those first thoughts and select a few for further progression. Once, however, product appearance becomes a key factor and you are taking critical decisions that affect appearance, then, even though this is only for your own reference, it is important to render each concept well enough to avoid misleading yourself. At some stage you may also need to discuss those ideas with your colleagues and while, as designers, they are visually literate, you need to give them as clear a picture as possible – your best ideas could fall at the first post because you failed to put the case effectively. Finally, you will often present sketch ideas to the client – both as back-up to the final solutions and in their own right. As back-up, they lead the client through your thinking and underscore the final recommendations and, if they are good enough to stand alone, they can be used to underline the conceptual nature of an early design phase. In these circumstances, it is too early to give the client a fixed idea of what the product will be like, and a tight, finished visual can erroneously give him the impression that the job is nearly over.

Too often one hears students excusing an inadequate drawing with the words 'it's only a sketch', believing that the word sketch somehow endorses a sloppy approach to drawing where mess is tolerable. The real secret to good sketch rendering, however, is *economy*: minimum marks for maximum information. In an ideal world this would work hand in hand with a loose, fluid approach which makes every sketch look as if it was dashed off in minutes, even though many designers actually work very hard at making their visuals look sketchy!

Example 1:
Mechanical Digger

This is a typical example of the kind of project where you might only be concerned with the design of certain key parts rather than every aspect of it. For example, the brief might be to redesign the cab but leave everything else unchanged, and in these circumstances you would almost certainly work from an existing underlay. Since you are not directly concerned with the design of everything else, you need to be able to block in these areas quickly, and sufficiently well, to allow you to concentrate on the job in hand.

Stages
1. The view was chosen from a photograph in a brochure and enlarged to a workable size (A2) leaving the cab area completely blank. Once the idea for this was sufficiently thought through, the whole drawing was traced out using a Nikko Finepoint and a slightly thicker Overhead Projection pen for the perimeter. The tread detail was also lightly sketched with the Finepoint so that it would remain visible under the Black marker. One way of rendering glass is to work from the back to the front; in other words, you draw something behind the glass that can be modified or distorted by reflections or tints. For this reason, all of the interior is blocked in, as if in silhouette, and a background is run behind the cab which can be seen through the glass.

2. A Cadmium Yellow marker is used first to block in all the yellow parts; the bucket has been done by streaking across its width with a lighter touch near the bottom. The upward-facing surfaces of the wheels are left untouched. Once this has dried, the same marker is used on all the right-facing surfaces to make them slightly darker. A Warm Grey 4 is used on the lights and hydraulic rams to establish the horizon and desertscape.

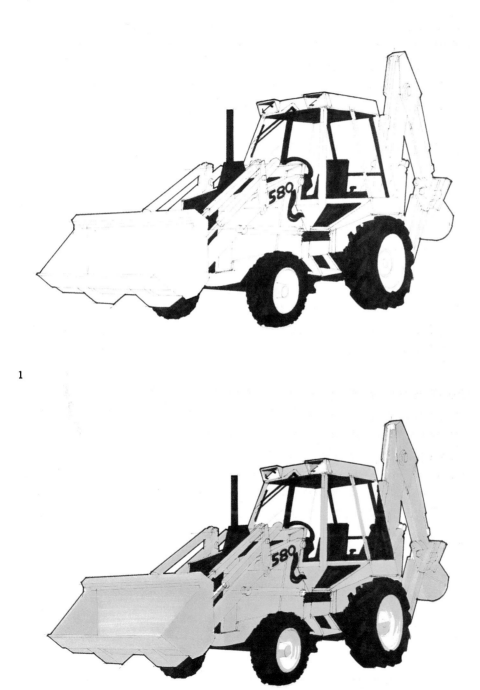

1

2

3. The background was masked up with black tape and then a Pale Blue marker used in a positive 'brush stroke' right across the tape, not worrying too much about maintaining a clean edge where the blue meets the yellow, because this can be cleaned up later. The Pale Blue is overlaid towards the right-hand side so that it is denser and, once the tape is removed, the ragged edge of the rectangle is tidied up.

3

4. To model the downward-facing surfaces further, a Yellow Ochre marker is used on the less dark parts, such as the undersides of the cylindrical parts, inside the wheels at the top and inside and under the bucket, and a Clay marker on the really dark parts, such as the wheel arch and underneath the bucket. A tiny bit of white pastel is put on the right-facing windows and taken off below the horizon level with a Cool Grey 3 marker. A white crayon is used to lighten the horizon on the side windows and to rough-model the exhaust and tyres (the white within the treads is removed with the Black marker). A sky-blue coloured pencil is used on the headlamps overlaid with fine black pen lines and white crayon. The black Overhead Projection pen is applied to near parts, such as on the bucket and the falling-away back edge of the cab, so that they stand forward from the parts behind. Finally, white gouache is used for the highlights.

4

Example 2:

Bottles

These ideas for blow-moulded cosmetic bottles would normally be drawn up individually as part of the design process but for the purposes of this demonstration they were done simultaneously. Where accurate proportion (to reflect volume) is important, it can be quicker and easier to work on elevations rather than on perspective views. If you do choose an elevation rather than a perspective view, it is just as important to be sure that you understand the shape from all around before you attempt to render it.

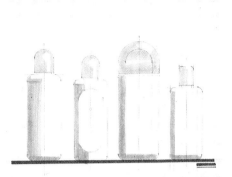

1

2

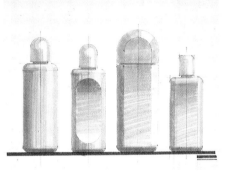

3

4

Stages

1. The form is sketched out using a Nikko Finepoint and the left-hand and lower edges shaded with a Cool Grey 4 (C4) marker. Pale Blue is laid in on the caps leaving a white highlight.

2. The three bottles on the right all have flat areas of plastic that should be tonally the same as the centre section of the cylindrical one. The right-hand side of the flat areas is masked up and a Cool Grey 3 (C3) wiped briskly across; this streaked effect promotes the impression of a flat surface. On the left-hand bottle the C3 and C4 are used together to model the cylindrical shape.

3. The darker areas are worked over with a C5 and C4 (again) to model the form further, particularly on the underside of the scallop on the second bottle. The left-hand side of the caps are worked again with the Pale Blue to provide the same degree of modelling as on the bodies.

4. Manganese Blue is used on the left and lower-facing surfaces to give more contrast to the modelling.

5. A fairly thick fineline marker, such as an Overhead Projection pen, is used to emphasize the left-hand edges and tidy them up a bit. The pink background is put in loosely behind the left edge to throw the images forward slightly. Finally a white crayon and gouache are used for highlights and details.

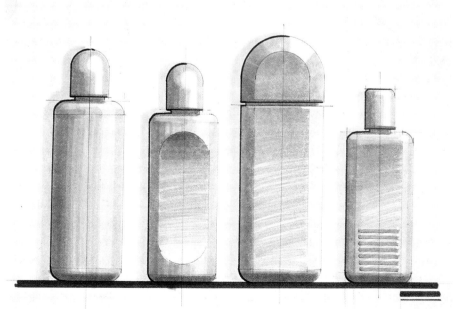

5

Example 3:

DIY Knives

This is a typical sketchpad visual where ideas are worked up quickly to see how they look. Normally, of course, only one drawing would be worked up at a time, but for the purposes of this demonstration, as with the bottles on the previous page, they were all done simultaneously.

Stages

1. All are drawn freehand using a black Nikko Finepoint and then further outlined along the lower edges using a thicker black pen.

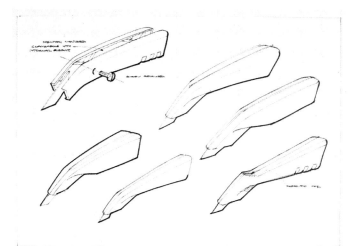

1

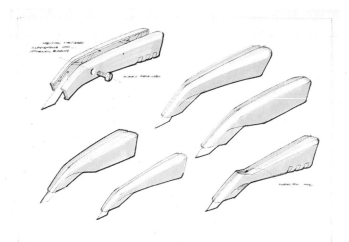

2

3

2. A Pale Yellow marker is used for the top surfaces, and a Cadmium Yellow for the sides, leaving a broad white highlight along the radius.

3. The lower surfaces are further darkened using a Yellow Ochre marker with a little bit of brown crayon in the deep shadow areas.

4. The blades are blocked in with a Cool Grey 4 and a shadow thrown across with a Cool Grey 6. To finish off the page, a yellow perimeter line is used to tie all the drawings together and provide a frame.

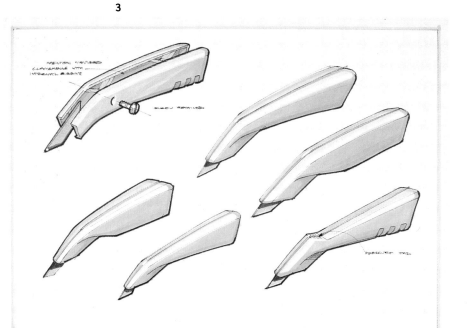

4

Finished Rendering

The design work is done, the solution agreed, and you need to produce a superb visual to show the client. Many designers reach this decisive point in the project where the creative work has to stop, because there is only sufficient time left for the finished rendering. The type of drawing you do will depend on your experience and ability, but you will probably be striving for as photographic a likeness as you can achieve, which also shows the product to its best advantage.

Example 4:
Vacuum Cleaner

A view looking slightly down on the cleaner was chosen to help establish the size of the product and to suggest that it is on the floor. The drawing was set up with the cleaner head and pipe lying in front so that a reflection of this can be seen in the side of the cleaner. There is also a reflection of the underside of the handle in the top surface which was constructed according to the method described on page 57. The cleaner was treated in the same way as the basic cube, with the top surface as the lightest and its nearest corner slightly darker to provide contrast to the highlight. The angled surface near the handle was made a little lighter because it would be reflecting something different. The horizon was run along the highlight and wrapped around the back.

Stages

1. The underlay is traced off using a red crayon (as this will not ruin the drawing if dislodged by the marker) and a fineliner used within the black areas to allow the details to show through. When tracing off, it is a good idea to put in simple registration marks (the cross by the wheel), just in case you get the underlay out of alignment and need to trace down a detail later.

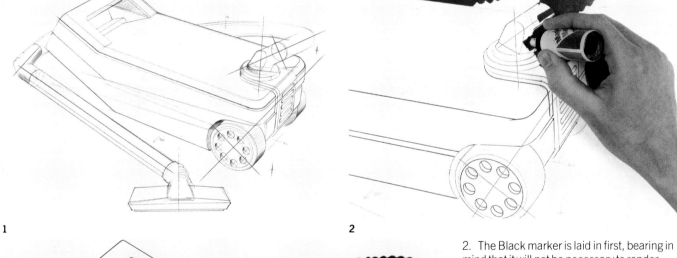

1

2

3

2. The Black marker is laid in first, bearing in mind that it will not be necessary to render every detail of the convoluted hose.

3. The Red marker stage is done next. Masking tape is put down along the angled lower edge so that the marker can be wiped straight over it. This is not strictly necessary if, as here, it is planned to cut out the drawing and remount it. However, it is good practice to avoid the build-up of mess because this can influence the way you see a drawing. For the same reason avoid testing markers on the sheet you are working on; use a scrap sheet instead.

A Cadmium Red marker is used as the base colour and then a Venetian Red for the darker areas and to suggest the tube's reflection. Next, the side is lightly streaked with the burnt orange colour of a Sanguine marker. Note that the reflection of the handle is also put in with marker.

4. The whole drawing is lightly dusted with talcum powder both to 'lubricate' the paper for the pastel, and to tone down the colours very slightly so that subsequent marker applications 'pull' the colours back (most noticeably on the black). Warm Grey 4 is used for the ground tones in the tube. Some red pastel is mixed up to match the Cadmium Red and applied to the top surface, working from the front to the back. Dark red pastel is added to the mix to darken further the near corner. Excess pastel is cleaned off with a roll of soft tissue which is revolved as it is used to avoid spreading the dust to unwanted areas. There is no need to worry about the dust getting everywhere, as, provided it isn't rubbed in too hard, it is easily erased and, in any case, hardly shows on the black and red areas. If two adjacent colours had to be put in with pastel, it would be better to mask off each in turn to prevent the spread of unwanted colour. Some designers favour the use of fixative before applying the second colour, but do not do this unless you are satisfied that a light application of fixative will not affect the surface and make more markering difficult.

4

5

6

5. To create the highlight in the radius, the pastel is removed with a clean soft rubber. The drawing is worked into again with the markers to remove talc/pastel and lift the colour. It is important to be selective or everything will return to how it was before; the blacks and alongside the horizon on the nearside radius are particularly emphasized.

6. Coloured crayons are employed to tighten up the drawing. The white is used on all the black areas to model their shape, and guides are used to keep the ellipses crisp. A blue pastel is used to create the sky tone in the tube and the horizon detail is sharpened with a black crayon running through to a sandy crayon as the tube wraps around. (Note how the reflection of the hose where it exits from the cleaner had been forgotten and had to be put in at this stage.)

7.   White gouache is used in a 'blobby' consistency to put in the highlights (the temptation to put one on every kink of the hose must be resisted).

8.   The drawing is trimmed out with a scalpel in readiness for the final mounting. An off-the-shelf graded paper was chosen as these look very professional. To give contrast to the lighter, left-hand end of the drawing, graded paper is laid down with the darker end to the left. Where there is a lot of untouched white paper, as here, it is better to cut away the background behind the rendering to prevent it showing through and dulling the colours. The rendering is placed on the background in the desired position and lightly marked around with a soft pencil. The background is trimmed away about 4mm inside this line so that there is no possibility of misalignment. Both the background and the drawing are stuck to a fresh sheet of paper before final mounting to card or foamcore.

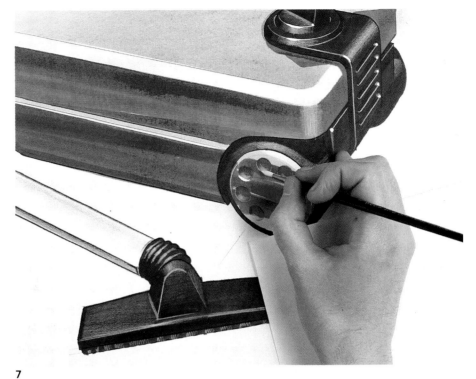

7

8

Example 5:
Electric Drill

If you were actually designing an electric drill
the chances are that, like the generator on
page 73, you would be working on scale
elevations and rendering them up as you go.
Your final presentation to the client would
probably consist of these visuals, backed up
by sketch models. However, for whatever
reason, you may not wish to make models at
such an early stage, preferring instead to
present a fully rendered three-quarter view
which can give a good all-round impression of
the product.

Stages

1. In this example, the underlay was drawn
out from an existing photo to establish the
perspective. The right view was found in a
catalogue and enlarged up to about life size
(either a Grant enlarger or a photocopier can
be used). To help keep the perspective true,
the minor axes of the main circles are laid in
first (chuck, handle-mount, motor housing)
and then the ellipses that define those shapes
are put in with ellipse guides.

2. The underlay is traced off onto a white
layout paper using a pencil and a black
fineliner for the black areas. All of the black is
put in completely flat so that an impression of
the drawing is immediately built up.

3. A Mid Blue marker is used for the blue
mouldings, with a streaky effect employed
towards the back of the top of the drill in the
hollow finger recess. The drill's shape is
treated in the same way as a cube with the top
surfaces slightly lighter, so, once the first coat
is dry, the side-facing surfaces are gone over
again with the Mid Blue. Antwerp Blue is used
in the darker recesses and under-facing
surfaces, and once again the streak effect is
exploited to model the curved surfaces.

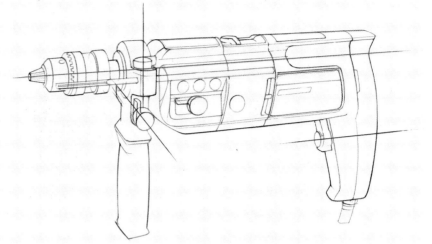

1

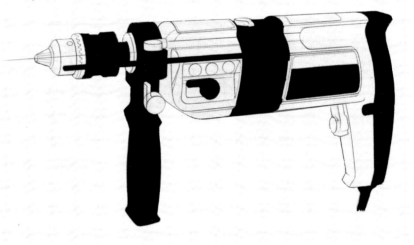

2

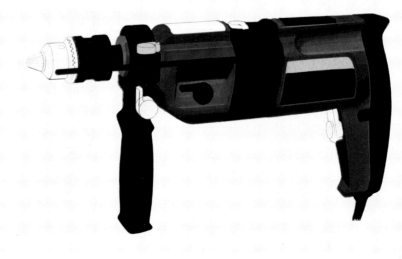

3

4

5

6

4. The yellow and red areas are put in flat and then the same markers are used again to produce crude modelling of the shapes. A Warm Grey 4 is used on the chuck as the basis for a desertscape. The whole drawing is lightly washed with talc to knock back the tones and then the Black marker is used again on the side-facing and lower parts to increase contrast, leaving the lighter (talced) parts untouched. The Antwerp Blue is used again to remove the talc on the lower-facing blue parts.

5. This shows the drawing at a stage about halfway through the pencil work with the white crayon used extensively on the black areas to model the radii into soft highlights. Note how the white has been worked across both the blue and the black together and then differentiated later with a black and a white crayon along the joint-lines. When doing joint-lines always use a black, or dark tone of the base colour, on the falling-away edge, and a white crayon on the facing edge. The desertscape is finished off with a blue crayon for the sky, a black for the contrasty horizon and a sandy crayon for the lower half. At this stage it was considered that the drawing lacked contrast, so a Prussian Blue was added to the deep recesses and under-facing areas. The red parts are modelled with dark reddy-brown and white crayons.

6. The rendering was now almost ready for mounting, with only the lettering to do. At this point the depth stop did not appear sufficiently in front of the drill; it really needed a shadow thrown across the main body of the drill, but this seemed a bit risky to do at so late a stage in the drawing. As a compromise, a slightly thicker black pen was used to outline the depth stop and handle to make them stand out from the main body.

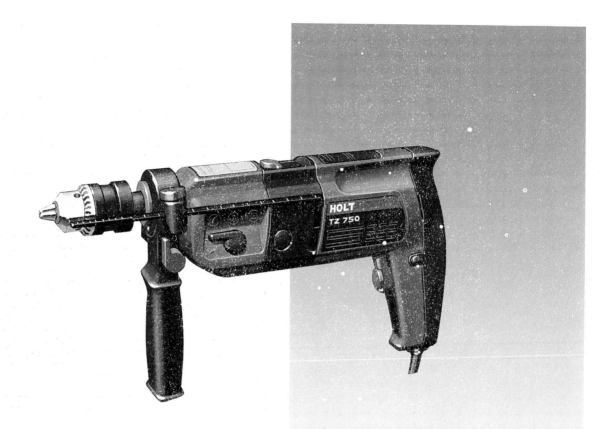

7

7. The finished rendering. White gouache has been used for the highlights on all the near corners and radii. The words HOLT and TZ 750 have been hand-lettered with white gouache, while the remaining graphics have been suggested by double parallel lines done with a white crayon. For final presentation the drawing is trimmed out and mounted to a graded background, so that it appears to float.

Example 6:
Generator

There are advantages and disadvantages to working on full-size elevational views. It is very easy, particularly for the inexperienced designer, to take liberties with the form and ignore the more complex three-dimensional problems of the design. On the positive side, however, you get a true sense of scale and therefore of proportion. With this kind of design problem you will probably be working both with sketches of three-quarter views, and with full-size technical drawings, especially if you are juggling internal components to achieve the best layout. Once you have a possible technical solution, it is a fast and effective transition to the fully rendered side-view. In our studio we would normally present a card or foam 'space model' (sketch model) to accompany the visual because many clients find it hard to read scale views. Nonetheless, a visual of this type not only gives the client a good impression of appearance, but also a genuine indication of scale and proportion. To help establish scale further, the drawing can be mounted to foamcore, trimmed out and photographed propped up in its intended environment, so that it appears real.

Stages

1. A good quality white layout paper that is unaffected by successive applications of tape and masking film was chosen and the elevational view was traced off using a black fineliner. The plan-view shown here is not a true plan-view of the entire generator but only of the top strip detail.

2. A Black marker is used to put in the fine details such as around the name plate and up the sides. Next, the black areas are well masked so that there is no possibility of inadvertently spreading the black to unwanted areas. Only two of the buttons were masked, using film, as the others will be black and can be put in later with crayons. In retrospect, it would have been quicker to ignore these two coloured buttons and render them completely separately for pasting up at the end. Black photographic tape is used to mask the radius along the top of the plan-view and down the right-hand side. A broad marker is made up (see p. 16) and the Black or Cool Grey ink is wiped from the bottom left to the top right in bold diagonal sweeps, using the marker in such a way as to leave the right-hand corner almost white. After a single pass, the tape covering the radii is removed and the process repeated, taking care not to overwork the area.

1

2

3. All the masking is carefully peeled off, revealing bleed under the tape; this is not serious because on a drawing of this size it really won't be noticed that much. The red front-elevation is masked out with tape, and film used to protect the name plate and the work just done. As with the top view, tape is used to mask a strip along the top and right-hand radii so that these will remain slightly lighter. Conversely the bottom and left-hand radii are masked up so that these can be given an extra coat of red. Once this is done, the broad marker is used in the same way as with the Black leaving the top right nearly white; the masking on the radii is removed and the process repeated. (For the purposes of photography the main red area of this drawing had six coats instead of two or three, so much of the contrast was lost – be sure not to let this happen. If you do overwork the inking, you will find that the colour pools and goes slightly sticky. This can be removed either by using a putty rubber, or dusting lightly with talc).

4. All the masking is peeled off and a Venetian Red and a Light Mahogany marker used to lay in the dark areas of the red parts and a Black marker for the darker areas on the plan-view (for example, the filler-cap recess) and control panels. Warm Grey 4 is streaked right across the middle of the lettering to create a horizon, and a Mid Blue and Cadmium Yellow are used for the buttons.

3

4

5. The drawing is finished off using the white crayon for all the edges facing upwards or to the right, and a dark crayon for all edges facing to the left or downwards. The desertscape is modelled in the usual way and white gouache used for the highlights at about two o'clock on the convex curves and seven o'clock on the concave curves (e.g. the button recesses). Use dry-transfer lettering if absolute accuracy is important or a white crayon if it is not. A medium-tipped black fineliner is used around the perimeter before trimming out and remounting to white board.

Example 7:
Motorbike

The underlay for this drawing was taken straight from a visual done for Yamaha in 1982. The brief was to produce themes and broad concepts for a range of bikes based on existing running gear. With this type of brief it is clearly a waste of time to draw all of the 'given' parts laboriously every time you develop a new idea; you only need to render them roughly so that there is sufficient information for the eye to fill in the necessary detail. Speed is important because you will need to turn over ideas quickly, but at some stage you may decide that, as here, one concept is worth working up to a slightly better finish for eventual client presentation.

1

Stages

1. The underlay has been traced off freehand (with the exception of the wheels which have been drawn with a circle guide) using a fineline black pen. All the blacks are laid in first to give a good impression of the whole bike.

2. The bodywork is put in using a series of Cool Greys modelled from dark to light and maintaining a horizon along the curve of the tank and the under-fairing. The discs have been treated with the characteristic 'turned' finish so beloved of hi-fi designers during the early seventies. (It is useful to understand why you get this effect on turned parts. Imagine the surface enlarged, like the grooves of a long playing record. Each segment of each concentric circle will reflect the same information but spread over a smaller length of its arc as the circles decrease in size. In other words, a light at about two o'clock would show as a highlight on the upward slope of each groove and a shadow on the downward side.)
    Steel marker is added to the Greys to give a slight blueness to the colour, and Warm Grey to the discs to suggest ground tones; the front indicator and rear lamp are also put in at this stage along with a Steel tone to the screen. Note how a white edge is left along the leading edge of the fairing to suggest the wrap-around of the screen.

3. The background is put in at this stage to throw the bike forward and define those areas where you see right through the machine. I chose a strong diagonal to match the chassis diagonal and this was done by simply placing my PVC straight-edge on the paper, drawing around it with the Cadmium Red and then filling it in.

2

3

4. White pencil is used to model the black parts and then the Black marker is used again to cut back into the white and create darker tones. The general form of the engine has been suggested with the white crayon, without attempting to give an accurate representation.

5. Looking at the drawing at this stage, I felt it needed more contrast in the bodywork so reapplied a darker Cool Grey in the deep shadow areas. The whole drawing was then finished off with white gouache for the highlights and outlined with a black pen. There is a real temptation with this kind of sketch to overtighten it so that it loses any sense of liveliness. If you are in any doubt about when to leave well alone, put the drawing aside for a time and work on something else; when you return to it you will find the refreshed view makes a decision easier.

4

5

Computers and calculators
by the Sharp Corporation, Japan

This set of four visuals was done by designers at Sharp in Japan and show two preliminary design sketches for an electronic calculator with printer (1) and a personal computer TV (2) and also two finished visuals for an electronic calculator (3) and a portable computer (4). The broad, bold wipes with an extra-wide grey marker give the visuals a nice fluid feel, and it is a technique very popular in Japan. There is plenty of contrast in all of them and the overall result is informative and punchy.

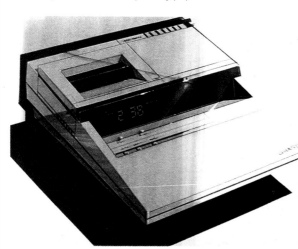

1

2

3

4

The renderings on pages 79-83 (top) were all done for General Electric Plastics for use in their promotional calendars and are therefore more like finished artwork than presentation visuals. The brief from General Electric was to come up with ideas, work up those ideas to a point where they can be understood and then do a finished visual. As with much conceptual work of this type, it is important to make the product look believable, even though, as products, they are not at all resolved.

Central-heating pump

This concept was drawn to show off the high finish available from Lexan polycarbonate, and this was achieved by using a strong contrasty horizon. Note how the reflection of the pipes onto the main body, and the semi-circle onto the key pad, help to make the mouldings look glossy. The metal casting in the middle is a semi-matt die-casting and therefore has less contrast along the horizon. The key pad in the foreground was done separately, by first using a Black marker and

then rendering in the blue buttons and steel-coloured LCDs before wiping white pastel across the whole face from top left to bottom right. This was then wiped away boldly with an eraser to produce the impression of a reflection; the blue grid over the surface was added with a coloured pencil. Despite the fact that this is not a true front elevation, the dry-transfer lettering works effectively.

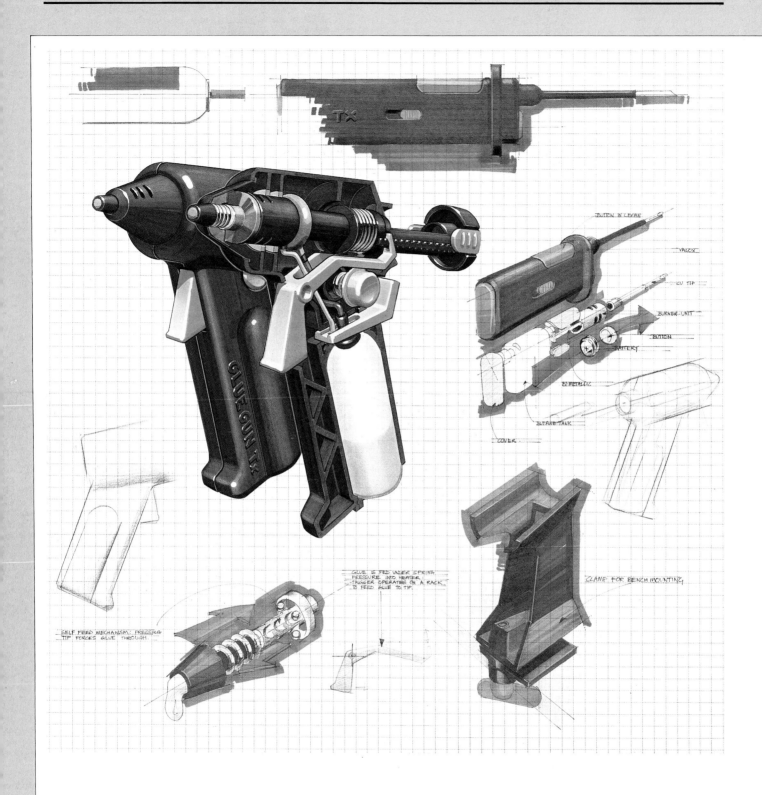

The following labels appear within the illustration:

BUTTON IN LEVER

VALVE

CU TIP

BURNER-UNIT

BUTTON

BATTERY

BI-METALLIC

BUTANE TANK

COVER

CLAMP FOR BENCH MOUNTING

GLUE IS FED UNDER SPRING PRESSURE INTO HEATER. TRIGGER OPERATES ON A RACK TO FEED GLUE TO TIP.

SELF FEED MECHANISM : PRESSING TIP FORCES GLUE THROUGH.

GLUE GUN TX

TX

Glue-gun and soldering iron

The intention was to make the whole drawing look like a page from a designer's sketchpad (of course, no designer actually works like this!). The main cut-away drawing was done first and then the underlay moved slightly to the left for the second view with the perspective 'fudged' to make it look right.

These two main views were done with markers and also some pastel for the orange upward-facing surfaces. The black has been extensively modelled with a white crayon so that it looks suitably matt; note also the orange shadow over the trigger in the second drawing. All the background sketches were done freehand in the same way, but slightly

less finished. To make them recede with respect to the two main views, the entire background was brushed with white pastel before the main pair were stuck down.

Motorcycle

The main three-quarter view was done first and the underlay shown to General Electric for approval before going on to the finished rendering. All the orangey-red parts were done with a combination of Cadmium Orange, Vermilion and Cadmium Red markers. The Cool Grey highlight on the tyres is not true to life but adds form to the shape. All three drawings were trimmed out and glued to the streaked blue background which was done using the felts from four markers (Pale Blue, Mid Blue, Antwerp Blue and Prussian Blue). I chose a non-absorbent Frisk CS10 paper because of the translucent quality you get with the marker ink;

unfortunately, it leaves a sticky finish which had to be removed by dusting with talc. The background had to be cut away behind the bike because it showed through too strongly in the white areas – a laborious process, particularly around the holes in the wheels.

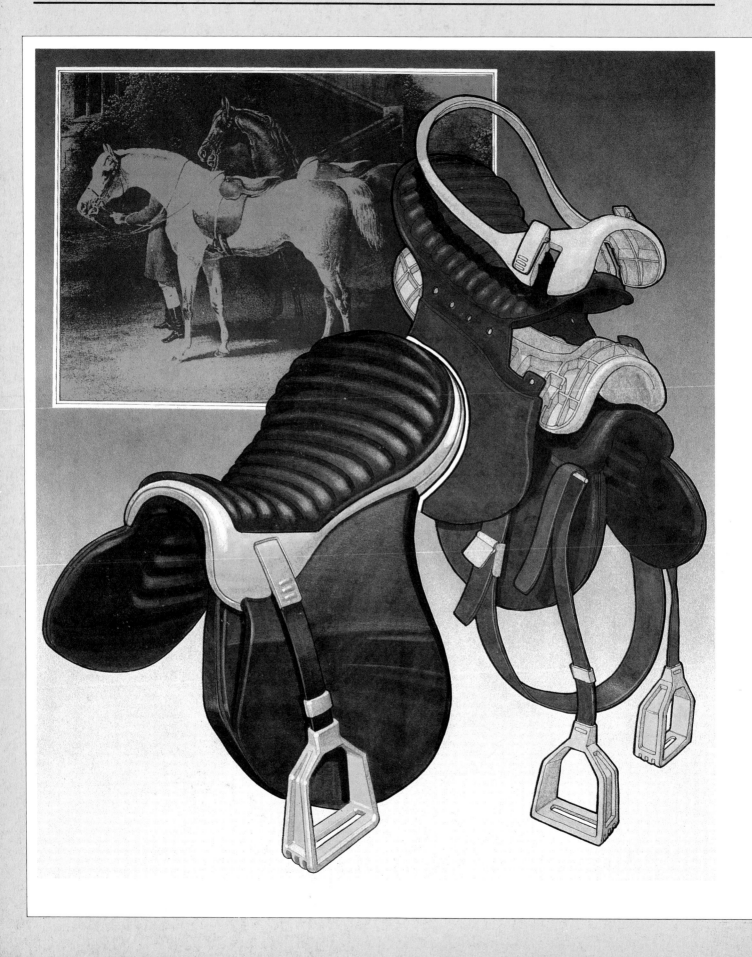

Left: Plastic saddle

The main drawing was done using markers while the background exploded view was inked first with technical pens before colouring up. The old etching of a pair of horses was photocopied on to the back of a piece of polyester drafting film and all three images stuck down to the airbrushed background.

Right: Motorized bicycle wheel
by Seymour/Powell

This drawing was done with markers and some coloured crayon, particularly on the wheel itself. The three background windows are all covered with Polyester film to reduce their strength with respect to the main drawing.

Right: Pedal car

This visual was one of thirty-six drawings presented to the client. The background window formed a linking theme across all thirty-six and the image in each window was varied to suit each product – in this case a desert. The window effect makes the jeep look as if it is jumping out of the frame and transforms the three-quarter view into a more interesting composition.

Above: Motorcycle controls
For Yamaha Motor N.V.

This visual was done for Yamaha to put across an aggressive sporty image to instrumentation and controls. Only the left-hand side of the bars is fully rendered to help give the feeling that this is a sketch. The really successful part of the drawing is the use of the bold reflection of the left-hand bar in the tank.

Right: Haircurler set
For Clairol Appliances Ltd.

This is one of a series of sketches done for Clairol, exploring alternative shapes for the moulded housing. It shows how a bold reflection can be used effectively when working up ideas quickly.

Left: Vacuum flask
For Fether and Partners on behalf of
Thermos Ltd.

*This visual of a vacuum flask illustrates well
the problem of rendering chrome, particularly
a vertical tube. To help give the impression of
a mirror finish it is essential to use what is
already available in the product, in this case
the reflection of the inside of the handle in the
main cylinder, and the reflection of the black
top on the top surface of the handle.
Reflections like this are difficult to predict so I
folded up some card and offered it up to a
piece of Melanex (mirror-finished polyester)
wrapped around a cardboard tube to give me
the necessary information. I used a strong
desertscape on the radius of the handle and
the suggestion of a table edge, or square mat,
in the base.*

Below: Portable tool-storage unit
For Fether and Partners

*In this example the client specifically wanted
an 'illustrative' visual, drawn from an accurate
General Arrangement drawing. I drew the
perspective for the closed version using the
cube technique described on page 28. To
construct the second view I used the
diagonals from the first to set up a new set of
vanishing points; this ensured absolute
dimensional accuracy. The marker
techniques are the same as the other
examples but note especially the deep
shadow within the box which changes as it
passes from red to grey to green.*

6 Airbrush Rendering

Airbrush rendering no longer enjoys the popularity among designers which it had before the advent of markers in the late sixties, simply because of the sheer sweat of doing an airbrush rendering, which can take almost three times as long as an equivalent marker drawing. For the designer, airbrushing is also a great deal less 'interactive' than marker rendering which is more integrated into the design process; with airbrushing it is very much a case of reaching a design decision, clearing the desk, planning the drawing and then executing it. Once this process is underway there is very little room for manoeuvre or last-minute changes to the drawing – you are committed from the minute you start masking.

Despite these drawbacks, the airbrush gives the finest finish and slickest effect of all the techniques available to the product designer and is certainly the most enjoyable and rewarding to use. Indeed, many designers take up airbrushing as a hobby for the sheer pleasure of doing it, and welcome the infrequent opportunity of using it as part of a presentation.

If you are a beginner, don't rush out and buy an airbrush in the misguided belief that it will, as if by magic, transform the quality of your drawings. Successfully mastering the finesse of control required is only half the battle – the real skill is needed for masking and planning the drawing and, of course, knowing exactly where to put which tone. If you are seriously interested in airbrushing and haven't already done so, buy one of the books which specialize in the subject because this chapter cannot hope to cover every aspect of it and, apart from the following section on basic techniques, is really intended for those who are already familiar with airbrushing.

Gouache is still the most popular airbrush medium among product designers and is very forgiving because mistakes can be easily covered by the opacity of the paint. Despite this, I prefer transparent inks because they require less masking, although greater skill is required to judge tonal balance. In addition, the colours have more depth and brilliance because the underlying white substrate can be seen through the ink. The technique with these inks is to work from dark to light, so, if you were rendering a matt cube, you would first peel off the mask from one side and give this a flat tone with the airbrush. Next, you would peel the masking from the second side and give this its first coat and, in the process, the previous, still-exposed side, its second coat. Finally, you would peel off the top surface and give this a light coat, and in the process give the two previously coloured areas a second and third coat respectively. As you will see, each successive coat makes the previously applied ones one tone darker.

The problems begin when you have a complex shape with maybe eight different shades of the same colour, and it is obviously very important to pull off the right bit of masking at the right time. Additionally, of course, it is also vital that you get the very first coat right, because there is a strong tendency to start too dark so that, when you are half way through, the ink has been overlaid as much as is feasible and the dark areas are already too dense, or, worse, become sticky and prone to lifting off with overmasking. Indeed, with all airbrushing, because you cannot see the whole drawing which is covered with masking, there is a tendency to do everything too dark – the golden rule, therefore, is to spray considerably less than you judge is right, at least until you have gained some experience. The great advantage with transparent inks is that you rarely have to reposition a mask once it is removed (and risk misalignment or ink getting underneath) and you only have to remask between colours, all of which adds up to considerable time saving.

Basic Techniques

This short section is intended for those designers who are picking up an airbrush for the first time and wonder where to begin. What follows will help you get started, but cannot be as comprehensive as some of the specialized books or, better still, tuition and demonstration by an experienced user.

Confidence. More than any other medium, you need to practise and perfect the basic skills of using an airbrush before you rush headlong into a finished rendering. Take the time to build confidence in your own ability by practising the basic exercises so that you no longer have to think about how to do them, only where to use them.

Getting going. Practise controlling the mechanism of the airbrush to give you the right balance of paint and air. For filling in large areas you will need all the air and all the paint that your brush can deliver, whereas for super-fine work you will need a small amount of paint and a correspondingly small amount of air. To avoid blodging you will quickly realize the importance of 'air on' first (before the paint) when starting to spray, and 'air off' last when stopping.

Line. Once you have familiarized yourself with controlling the mechanism, practise some freehand lines, trying to achieve an extra thin one (such as the example which is often included with the brush as part of the quality control testing) right up to a broad swath of colour. You will find that width is dependent on distance from the paper – working close to the paper will produce a fine line, and working far away from it will spray a wide one. You will also find that it is not *that* simple! As you get closer you need to back off the 'throttle' and reduce both paint and air to avoid producing a blodge; you will also need to combine this delicate throttle control with the speed of your brush across the paper. In other words, it is alright to throttle up when working close to the paper provided you are moving the brush fairly speedily, but if you slow down, or stop, you will quickly overload or flood the area directly under the brush. Once you are fairly confident of dialling up line widths to order, practise working with a ruler to produce straight lines.

The top line in this example of airbrushing different line widths (as shown in the sequence below) is done with the tip of the nozzle about 10mm above the paper and using a bridge to guide the airbrush in a straight line.

The second line is the thinnest achievable. The nozzle of the airbrush is actually placed on the paper and up against a ruler or bridge but, of course, off to one side of the area to be sprayed. This 'start-up area' is used to check the throttle settings for the fine line you intend to spray. The airbrush is then slid along the ruler edge to produce a straight line.

The third line is done in exactly the same way but with the bridge tilted up to raise the nozzle about 20mm from the paper.

The fourth line, or swath, is done by tilting the bridge right back and overspraying several times.

The fifth line, a very broad swath, is done freehand by passing the airbrush repeatedly across the ground to build up the density.

The final example shows the ruler used as a mask to create a sharply defined edge: the ruler is held firm to the ground and the ink sprayed right into the edge from the paper side. Spraying from the other side, over the top of the ruler, will produce a slightly more hazy finish.

1

2

3

4

5

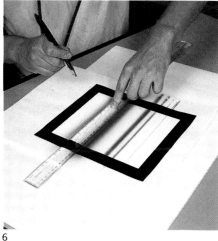

6

Tone. Producing a completely flat area of colour with an airbrush is surprisingly difficult to do, but well worth practising. Perhaps more appropriate to the medium is the graduated tone which grades from very dark to very light – practise it on very small areas and on very large areas (such as backgrounds).

Freehanding. This is the term I use to describe any airbrushing where spraying starts and stops within a masked area. Most of the time you always start airbrushing off to one side of the area to be sprayed (usually on the mask) and then move the airbrush over the intended area – all without really thinking about it. The masking itself provides the edge to the area you are working on, while you control the tonal balance and effect within the masked area. Often, however, you will need to produce a soft, fuzzy shape or line that is not bounded by any masking, which would of course leave a line. A good practice exercise to test your skills is to try and airbrush a sphere without any masking at all – no starting and stopping areas and no edges.

Masking up. The first difficulty that most beginners face is actually laying down masking film, particularly in large sheets, so that there are no air bubbles or uneven patches. The two most usual problems are getting the sheet down, finding that you misjudged it and haven't covered the drawing, and manhandling a large sheet into position without it folding up on itself. The adjoining illustrations show how to avoid these problems.

Cube, cylinder and sphere. Start by trying to render our old friends in a matt or satin finish; you will quickly find that it is possible to produce startlingly good results very quickly. Progress from these to glossy versions, and then into more complex shapes – a good one to practise is a smooth, doughnut shape because the mask can be cut exceptionally quickly with two passes of a compass-mounted scalpel blade, and it also demands deft control and freehanding.

General preparations. As with markers, make sure that you have everything to hand and that your hands and work area are clean. If you are using a board as your ground, wipe it over with a tissue soaked in lighter-fluid to remove any greasy fingermarks that would not show until you tried to overspray them. Apart from the inks themselves you will need two water containers (one for clean and one for dirty water), a roll of tissue paper for cleaning out between colours, a large brush for loading colours into the airbrush, and a

dustbin to spray into when rinsing through. Finally, make sure that you know exactly what you are going to do and in what order – if you are unsure, experiment on a photocopy or sketch *first* before you get too deep into the drawing.

Masking up

1. Select, or cut, a sheet of film sufficient to cover the drawing and lay it face down (i.e. film side toward you) on the drawing. Slide it to one side and roll back about 50mm (2in) of film and trim off an equal amount of the backing sheet.

2. Still holding the rolled-back portion, slide the sheet over the drawing until the other edge is well over the area to be sprayed; gently lower the rolled edge onto the drawing surface. This positions the film in the right place and ensures that no area will remain unmasked.

3. Lift the sheet at the unexposed end and fold it back on itself to free the cut end of the backing sheet so that it can be slid out from underneath. Once you have a hold of this edge, fold the whole sheet back again so that it is lying over your hand. Using a soft brush, or cotton wool, clear the previously rubbed-down 50mm strip of air-bubbles by wiping them out to the sides. When this is done, slowly pull away the backing sheet, wiping continuously as you do so with the other hand to remove bubbles and wrinkles.

It sounds complicated but really isn't, and will save you a lot of time.

Example 1:
Electronic Movie Camera

The underlay for this camera was taken from that done on pages 42-3. This in turn was originally done as an entry for a 'Camera of the Future' competition and, to give myself the best chance of winning, I decided on an airbrush presentation to seduce the judges (successfully as it turned out!). This was not the first camera I had done, and had found airbrushing to be the only way to simulate the complex coating found on lenses. Also, of course, it works wonderfully well on cylindrical parts, as here.

I planned to use only three masks – one for the blacks, one for the bluey-greys and one for the small parts (buttons, lens etc), but things didn't turn out quite as intended. The cylindrical parts have a soft highlight running at about two o'clock on the concave surfaces and seven o'clock on the convex ones. The body parts are treated in the same way as the basic cube with the nearside, angled face reflecting the most light (to match the two o'clock highlight on the cylinders).

Stages
1. The back of the underlay is coated with graphite, using a 6B pencil, and this is rubbed in gently with some cotton wool; a 2H pencil is then used to trace the underlay through onto a sheet of 'Line and Wash' board. A sheet of layout paper sufficient to cover the entire board is chosen and a window cut in it to expose the line drawing. A sheet of masking film is laid over this using the technique described on the previous page and the perimeter of the black parts and each individual crescent-shaped segment are carefully cut out.

The forward-facing *rim* of the lens and the rangefinder (far side) are peeled off and sprayed flat and then the tiny part of the rangefinder visible beyond the lens is peeled off and sprayed. Next, the furthest of the two crescents inside the lens hood is peeled off and sprayed to a soft highlight. This mask is replaced and the process repeated for the nearer of the two, again replacing the mask when it is finished. The nearest of the concave sections on the hood is peeled off and sprayed to a soft highlight at two o'clock, the mask replaced and the next two sections removed. These can be done together because they are approximately the same diameter and there is no need to define an edge (because they are surrounded by a different colour). It is not necessary to replace these masks after spraying because there is no immediate danger of overspraying; they are simply covered with a sheet of paper. The process is repeated for the eye cup, replacing each section as the next is done, with, again, no need to replace the final one.

The microphone is divided into four areas: the highlight itself, running around the top and down to about three o'clock, the continuation of the highlight running on round the radius that divides the front face from the cylinder, the cylindrical section and the flat front. The highlight continuation is peeled off and a wisp of black blown across it. Then the cylindrical section is peeled off and the form modelled to the same two o'clock, soft highlight as the lens hood. Care is taken not to overdo this, because it is only done to differentiate the various parts from each other. Next, the front face is peeled off and a graded tone given from top to bottom, concentrating the spraying in the lower half where the tones should nearly blend. Finally the mask is peeled from the highlight, which is sprayed from the far side round to about one o'clock to grade the radius.

The small bit of protruding cable was done before the grommet and then the underside of each convolution of the grommet lightly sprayed using a small coin as a mask; this was overlaid again with a simple tubular effect running to a soft highlight up the middle. All of the masking can now be removed.

2. A piece of masking film is relaid across the drawing and all the grey areas trimmed out. The masking is peeled from the darkest parts first – the deep shadow beyond the microphone and the lower-facing radii – which are sprayed. Next, the masking is removed from the slightly angled lower-facing edge beneath the lens hood, and this is sprayed. All the masking from the forward-facing surfaces, including that of the lens-mount bezel and the edge of the arrowed groove, is removed and these are sprayed flat.

1

2

3

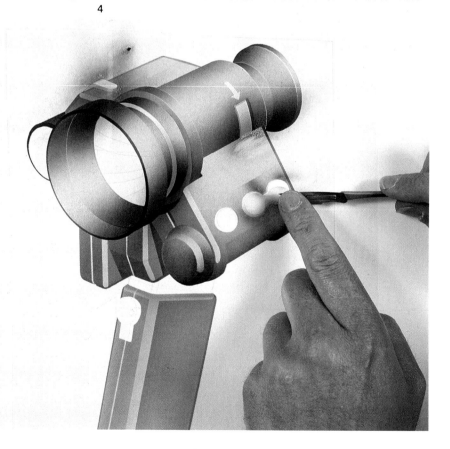

4

3. Next, the cylindrical part of the main body is airbrushed to the same soft highlight as the black parts done earlier. To avoid the cylindrical effect overlaying onto the forward-facing rim of the bezel, a tracing paper mask is cut to protect it.

4. This shows all the grey work complete. After the cylindrical body, the handle is airbrushed by doing the falling-away shadows first and then the left-facing surface, followed by the forward- (and slightly upward-) facing surface, leaving the highlight and moulding joint-line covered; note how the surfaces are graded from top to bottom. Finally, the radii joint-lines are uncovered and lightly oversprayed to take the edge off the white from the top downwards. After the handle is done, the final spraying is carried out on the angled diagonal face and the top surface beyond the cylindrical unit. All the remaining masking is carefully peeled off to see how successfully the tonal balance has been judged.

5. The drawing is remasked and all the remaining coloured parts are trimmed out ready for spraying. The button on the top surface is done first with the masking cut away around the vertical face, this is

5

airbrushed in vertical passes leaving a highlight at the nearest point. The mask on the top surface is removed and a light tone lightly washed on. This completed button is covered with a sheet of paper and taped along the nearside edge to prevent any possibility of overspray.

Next, the rectangular panel beneath the arrow is tackled: the mask is removed from the entire area and moved down to leave a small 'L'-shaped strip as a shadow area. This is lightly airbrushed, and then the masking is removed altogether, along with the masking covering the arrow and the blue band around the lens. These areas are all cylindrical and are treated in the same way, shading from the bottom up to the highlight at about two o'clock.

The protective paper is moved to cover these areas and work started on the double button: the masking is removed from the vertical edge, which is modelled to a highlight at about eleven o'clock, and then the masking is removed from the face. Because the buttons are slightly convex they need to be darker at the bottom than at the top, so the application is graded with the airbrush to achieve this. As before, the areas that have just been sprayed are protected. The trigger button is done next and is treated in the same way.

The airbrush is cleaned out and switched to red, before treating this button also in exactly the same way. (At this point I suddenly realized that I had completely forgotten the black band between the handle and the body so this was also sprayed in at this stage). The rangefinder cover is sprayed around the top with a light wash of sky blue to make it look silvery. The lens is also done at this stage by successively peeling off strips and overlaying orangey-browns. When faced with a problem like this it is important to get hold of some visual reference to work from. When I first did a camera I sought out a good photograph from a brochure and then simply copied what I saw in the lens and this worked very well. When doing this demonstration I felt confident enough to do the lens from memory which was, in retrospect, a mistake because it took a lot of effort to get it to the stage you see here, which is, as a result, desperately overworked.

6

6. All the masking is removed again to look at the whole drawing. At this stage I considered that I had not put sufficient contrast into the drawing and decided to remask the black areas and darken them further. Excessive masking on oversprayed areas can cause tiny patches of paint to lift off with each new mask; the surface was showing signs of this happening, so I elected to use a paper mask for this work. To do this, a piece of tracing paper is simply laid on the drawing. The areas to be cut are drawn on it and then trimmed out one at a time. In this case the mask was retained in position by coins around the crucial edges (special care had to be taken not to let the spray lift the mask). I rolled back each segment, leaving the end still attached

to the sheet, and weighted it with another coin. This allowed all the lower-facing dark areas to be toned down a lot further, but extreme care had to be taken not to leave a tell-tale line at the point where the rolled-up segment leaves the surface of the drawing.

7. The final photograph shows the completed drawing with all the edges tidied up using crayons (a dark bluey-grey for the falling-away edges on the grey parts and a white for the highlight areas throughout). With the airbrush work, all the detail of the microphone had been lost but the master was kept firmly attached to the board throughout, so it was still in perfect register. The back of this one area was lightly coated with white pastel and the necessary detail traced through; the holes were blacked in with a fineline pen and the edges of each hole modelled to a highlight at about seven o'clock with the white pencil. Finally, white gouache was used to create highlights on all the near corners and radii.

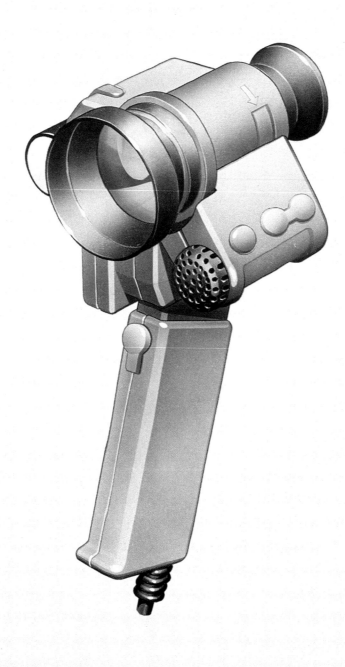

Example 2:
Telephone

This example shows how something very glossy – an office telephone – is airbrushed. As with the markers, the same principles of rendering apply and the form is treated more in terms of its reflections rather than the position of the light source.

Stages
1. Planning is very important with airbrushing, so a photocopy of the master underlay was taken and a detailed plan, using coloured pencil and marker, was worked out. On the green plastic area the reflection of the curly cable in the side of the telephone, and the reflection of the grommet where it enters the handset, are used to help suggest the glossy finish. All the left-facing surfaces are made slightly darker than those facing the front, and, rather than making the near corner of the top surface darker (as for example on the vacuum cleaner on p. 67), a lighter surface was preferred at the front, grading it darker towards the back.

2 and 3. A good line board (in this case Frisk CS10 which gives a very smooth finish) was chosen and the master underlay traced down using a hard pencil. A sheet of masking film is laid down ready for spraying all the green parts and, as with the camera, the remaining board is carefully covered with a sheet of layout paper. The perimeters of all the green shapes and each individual area of colour are cut out with a scalpel; the masks of the darkest areas, such as under the handset are peeled off first. The uncovered areas are sprayed with a light coating of green ink and then the next area is successively uncovered for overspraying, ending with the lightest (highlight areas). To keep everything in the right order, the rough was marked up with each area numbered in the right sequence. This also helps to avoid the problem of something being left out completely, which, if not spotted, can mean remasking just to do one bit.

Photo 2 shows the end of this stage just before complete removal of the masking. Note the pieces of masking already stuck around the edge: the sausage-shaped handset top with, below and to the left, the 'Y'-shaped mask from the near corner of the handset. Note also how difficult it is to read the drawing under the masking when all the masking is still in place. Photo 3 is at exactly the same stage but with all the masking removed.

1

2

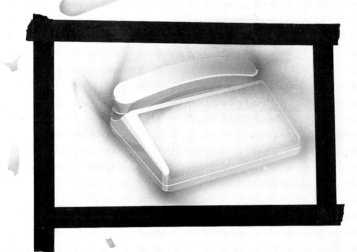

3

4 and 5. A clean sheet of masking is laid over in readiness for spraying the dark grey areas. As with the green, all the left-facing areas are made slightly darker than the front-facing surfaces, with the angled ones lighter still and graded from the near to the far corner. The darkest areas are the button recesses and so these are done first, followed by the downward-looking faces of each button and the rim above the LCD. Photo 4 shows the masking being removed from the left-facing surfaces, leaving the highlight and top surfaces for later passes. Photo 5 shows everything being oversprayed with the highlight masking just removed and the tops still covered.

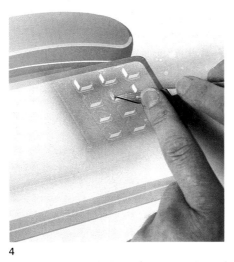
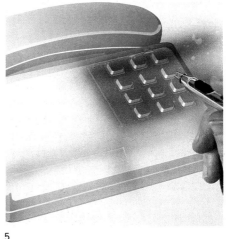

4   5

6. The end of the dark grey spraying reveals considerable mess. The coiled cable has been done in the following economical fashion: the masking is removed from each coil in turn and that area sprayed to a highlight, not as heavily as finally intended but just sufficient to define the edge between it and the next one (once defined, overspraying will not lose this edge). The masking is then removed from the next coil and the process repeated, lightly overspraying the first in the process. Working along the cable in this way means that three or four coils are done at a time and the tonal balance is maintained. Each coil can be further worked by using a curved paper mask to define the edge of each coil better. Note all the tiny bits of coil masking along the left-hand side.

6

7. This shows the drawing at exactly the same stage but without the masking. The grey has been rather overworked and is now too dark when compared with the green. Because of this, the button detail has been all but lost and will need putting back at the end with a crayon. The lack of contrast in the green necessitates a respray of these areas so, after masking, the green is gone over again, particularly in the deep shadow areas where a touch of dark blue in the ink establishes contrast.

8. The final mask is for the acrylic cover and the LCD. Often when you look at acrylic sheet the cut edges are substantially darker, so these are done first, fading them out towards the near corner. The reflection of light at the near corner will make it difficult to see through the acrylic at this point, whereas towards the back more will be seen through the surface. In this case there is only white paper under the cover, but if there had been, for example, a picture, it would have been almost totally obscured at the front and

7

almost totally visible at the back. The shadow on the LCD is sprayed first, before washing on a flat tone for the rest of the face.

9.   The final shot shows the drawing after some work with the white pencil, particularly around the buttons, and with a dark pencil on the falling-away edges. White gouache is used for highlights on all the near corners.

8

Example 3:
Cutlery

The airbrush can be a very useful tool when working on side-elevations or plan-views, principally because the masks are invariably easier to cut. In this case the requirement is for a satin-finished, stainless steel cutlery set, where the subtlety of curvature would be difficult to do well with any other medium.

Stages
1. The underlay for the knife, fork and spoon was lifted straight from a dimensioned drawing and many photocopies of the fork and spoon (which you see here) taken to establish the best approach: the spoon is the most difficult because of its three-dimensional concave curvature. Those surfaces looking up or to the right (i.e. inside the spoon on the left and the bottom) will be reflecting something dark, while those looking down, or to the left, will reflect something light. Once the rough is settled, a sheet of masking film is laid onto a blank sheet of Frisk CS10 paper and the original is photocopied onto this 'sandwich'. This method completely by-passes the tracing-down stage and leaves no traces of pencil on the surface to be sprayed.

2. Using the scalpel, a sliver of masking is trimmed out along all the right-facing edges of all three, as these will be darkest, and they are lightly sprayed with a bluey-grey ink. Next, the dark corner just above the knife handle is trimmed out and sprayed. The masking is

1

removed from the rest of the blade with the exception of a thin sliver running up the left-facing edge which will be left white. The grey is graded very slightly from top to bottom as this area is given no more than a whiff of colour.

3. A sheet of protective paper is placed over the knife and fork while the spoon is tackled. As with the knife, all the masking is removed from the spoon with the exception of a sliver

up the left-hand side, and the dark tone is put in freehand being careful to keep the direction of the spraying action running across the spoon at the point where it 'necks' in to the handle. The spoon is then covered and the whole process repeated for the fork, bearing in mind that this is not a compound curve so that the portion near the neck is looking up (the page) and therefore dark, while the portion nearer the prongs is looking down and therefore light.

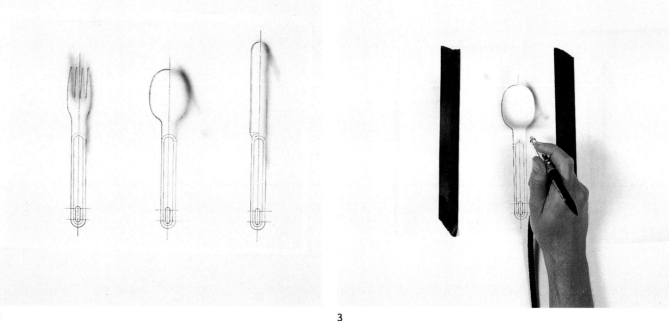

2   3

4. Since the drawing is on the film, and not on the paper, it cannot be removed and remasked. A strip of film is laid over the area where handle and metal meet, and a scalpel is used to cut lightly around the handle leaving the film on the previously unprotected area; this strip of film is joined with masking tape to a sheet of paper which protects the rest of the previously sprayed areas. The handles are all done simultaneously, starting with the right-hand upward-facing bevelled edges, including the ones inside each hole; these are graded where they wrap around the corners. The flat faces are done next, followed by the left and lower-facing bevels.

5. Once the spraying is complete the three views are carefully trimmed out with a scalpel, cutting around the unsprayed left-facing edges. They are then mounted down to a dark background to give contrast to this edge and the other white areas.

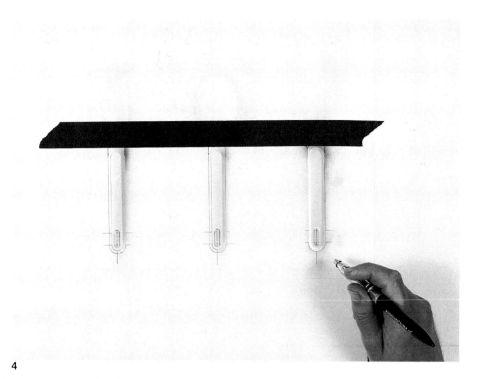

4

5

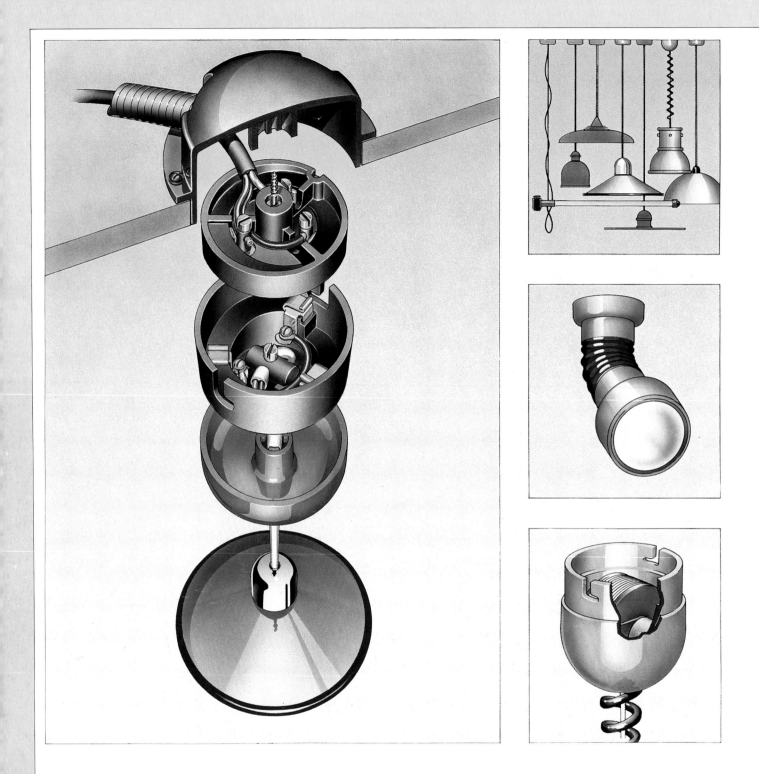

Light fitting

This explanatory visual was done for a General Electric Plastics calendar. After the idea had been roughed out and approved by the client, it was detailed up with a pencil. This was traced off onto tracing paper using technical pens, but only on falling-away edges (obviously we didn't want a black line around a highlight). Next it was enlarged photographically via an inter-negative and printed onto old-fashioned bromide paper, and then dry-mounted to CS10 board ready for airbrushing. Two prints were taken so that one could be used to practise on.

All the green areas were done with the first mask: note the crisp, well-defined highlights and the reflection of the boss in the flat surface of the inside of the cover. All the grey parts were tackled next, working from dark to light with minimal replacement of masking. Three further masks were used for the red, beige and yellow areas, before putting the airbrush aside and finishing off with gouache and crayons.

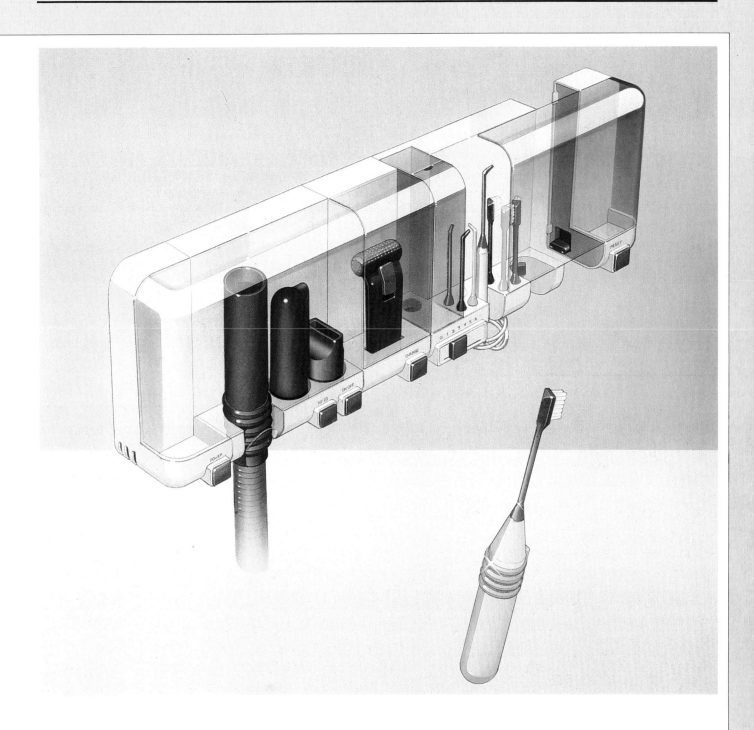

Bathroom unit

Also done for a General Electric Plastics calendar, this visual of a bathroom unit was one of those drawings which kept going wrong. I elected to use airbrush because of all the overlapping smoked polycarbonate covers, but regretted this decision upon removal of the final masking. This was an early acquaintance with translucent inks, and I was still working with the medium as if it were gouache, replacing masks as I worked, so I had no real idea of how the thing looked till the last minute. Removal of the mask revealed massive bleeding underneath and I had to set to with a brush the night before the deadline to rescue the situation. The blue background was put in to obscure much of the bleeding but it was too late to run this behind the 'exploded' cover on the right-hand side without risking everything. This example illustrates how a drawing can be rescued from what, at first sight, seems terminal disaster.

Right: Safety helmets
by Seymour/Powell

Commissioned for the 1985 General Electric Plastics calendar, this view uses airbrush for the helmet shells and marker for the rest, with careful inter-cutting between the two to achieve a good match. The underlay was built up from the yellow helmet and then reduced on a Grant enlarger for the two others. The shells were airbrushed first using Magic Color inks onto CS10 paper, and then the visor and liners were drawn onto marker paper and coloured up separately. All the elements were then assembled onto a new sheet of white paper and a whiff of ink sprayed onto the visor (either side of the highlight) to give it some presence. The background was set up and photographed, and the enlargement 'knocked back' with two layers of polyester film and one of K Trace.

Below: Modular process control system
by Don Tustin

The underlay was drawn out to scale and then transferred to water-colour board. The meters and buttons were masked out, and the dark-grey background panels airbrushed in with gouache. The latter were then remasked so that the doors could be sprayed darker. The shadows under the knobs, the beige meter recesses and the red graphic panels along the top were all airbrushed next. The silver back panel and shadows were also airbrushed, while the rest of the rendering was done by hand with a brush and dry-transfer lettering was used for the graphics.

Above: Infra-red optical reading head
by Don Tustin

This is a good example of rendering side-elevations. The red side panel was done first and slightly graded from the top, before remasking and putting in the horizon. This was then masked off with tape and paper so that the small cylindrical section could be airbrushed. The lettering was masked throughout this stage so that a separate

masking and spraying operation was unnecessary to create the white SM4. The lighter part of the black heatsink was sprayed next and a separate mask used for the darker shadows. The rest of the drawing was done by hand with crayons and a brush, and the whole drawing was remasked so that the grey background could be sprayed.

7 Coloured Paper Rendering

Using coloured paper as a base for rendering has rather gone out of fashion despite the advantages it can offer in some applications. It really comes into its own when rendering transparent materials, metals, jewellery, fabrics and furnishings, for quick and effective sketch rendering, and for objects which are largely of one colour.

The usual approach when working on coloured paper is to use the colour of the paper as the *mid-tone* value of the colour of the object, and then to darken the shadow areas with marker, pastel or crayon, and lighten in the light areas with pastel, crayon and paint. By using light media on the darker ground, high-contrast highlights can be produced which really make the drawing come alive. This also explains why the technique works well for metals and transparent materials which, of course, also tend to have high-contrast highlights. There is a further advantage in using coloured paper for transparent objects because, as discussed on page 138, the secret is to work from the back to the front. Objects appear transparent because we see the background behind them, and usually this background is modified in some way by the presence of the object in front of it. The coloured paper itself can form this background and the object can be sketched in with the minimum of marks using both light and dark pencils, and one or two markers.

Because of this economy, the technique works well for sketch rendering: ideas can be drawn straight on to the paper using a light pencil and modelled up very quickly. Since you are not working from light to dark (as you would be on white paper), you do not have to worry so much about the tonal balance of the final sketch: you are starting from a mid-tone and working either side of it. Of course, if you use a very dark paper then you will be working from the darker tones back towards the very light tones, such as in the car drawing on page 10, which can produce some stunning results but is, perhaps, slightly more difficult to do well.

The main disadvantage of coloured paper is that it is not transparent and so you can't simply slip the underlay underneath and trace it through, as you can with layout paper.

If you are not confident enough to sketch directly on the surface, or if the rendering is to be in any way a finished drawing, then you will have to trace the underlay down onto the coloured paper. This is done by rubbing the back of the underlay with an appropriate coloured pastel so that is evenly coated, and lightly brushing away excess dust; the underlay is then taped to the coloured paper and traced through using a fairly hard pencil. Alternatively, tracing-down papers, which work like carbon paper, are available in a variety of colours and are used between the underlay and the coloured paper.

Using coloured paper becomes less appropriate if the product you are drawing is made up of more than one basic colour; the odd flashes of colour in small quantities are of no consequence, as these can be put in with opaque paints such as gouache. Anything more, however, is a problem, because the base colour is so dominant that it effects any transparent media laid over it. For example, laying a Cadmium Red marker across a blue paper will only create a darker blue, or perhaps, a dark purple. This can be enlivened by overworking with a red crayon, but the result is never as strong as the original, bright Cadmium Red.

Both Ingres paper and Sugar paper have some 'tooth' (roughness) which makes pastel and particularly coloured crayons work extremely well. Because of their high rate of absorption, however, they drain the colour out of markers in no time at all, making it almost impossible to obtain a flat finish of any consistency; infilling of very large areas with marker, therefore, is very difficult to do and should be avoided where possible. If you allow the marker to dwell on the paper, it very quickly forms a darker patch which is difficult to blend in, although, if the marker colour is darker than the paper colour (such as with black), this is hardly noticeable.

Nevertheless, when using marker on these papers you need to plan where to end each stroke and how to make the streaking effect of the marker work for you, rather than against you.

Example:
Pencil Sharpener

The pencil sharpener is to be predominantly blue with a black front and a transparent cover beneath, making it a suitable candidate for rendering on coloured paper. The view is planned so that we see through the transparent cover to the base below, and out through the side to the background. Clearly the shape is identical to the basic cube and can be treated in the same way: the left-facing surface will be reflecting the ground off to the left, and the right-facing surface the ground off to the right. The top will be reflecting light from above, and so this will be lighter than the two sides. The back corner will be lighter than the near corner to add contrast to the bright highlight running around the nearside radii. Note the reflection of the handle in the upper surface (seen as an interruption of the general reflection in this surface), which, more than anything, adds to the impression of a glossy finish.

Stages
1. The basic underlay is drawn up using the cube method described in Chapter 3. Note how the centre of each face is located using diagonals so that the hole on the front face and the centre line for the handle can be exactly located on the same axis. The handle is then drawn from this centre-line upwards, and backwards, from the back face of the sharpener.

The exact position of the reflection is worked out according to the method described on page 57. First, the top face of the sharpener is extended horizontally until it meets the front vertical face of the handle. The points at which the lines describing this front face meet the extended top face are the points from which the lines describing the front face of the *reflected* image will spring. Next, a vertical is dropped from the centre of the top 45-degree ellipse of the handle; this line will bisect (in a perspective sense) the line where the extended top surface and the front face of the handle meet, and is extended on downwards. An equal distance downwards (in perspective) is marked off to locate the centre of the reflected ellipse, and a minor axis that shares the same vanishing point as its real counterpart is drawn through this point. The ellipse can then be drawn in, and tangential lines drawn to it from the previously defined points (where the handle meets the extended top face). All the unwanted construction lines are removed, particularly those defining the extended top face which will otherwise be confusing.

A light blue pastel straight from the box is used to coat the back of the underlay and rubbed in lightly with a tissue. The smooth side of the Ingres paper is used and the underlay taped down, before being traced through with a fairly hard pencil. Before the tracing has progressed too far, one corner is lifted to make sure that a satisfactory result is being obtained on the paper. The underlay is left taped to the Ingres paper along its top edge, in case it is needed again.

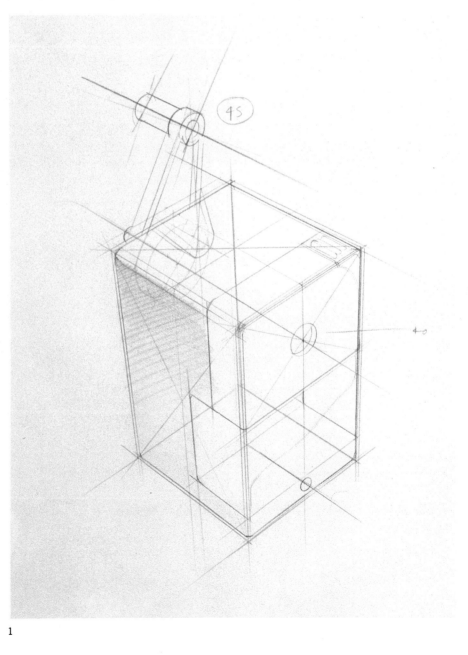

1

2. The marker work is tackled first: a Black is used for the front, without worrying if the previously traced-through line-work is totally lost, as this can be put back later. The left- and front-facing surfaces (including the handle and its reflection) are treated with a Mid Blue. The left-hand face is graded by using an Antwerp and a Prussian Blue towards the top where contrast is needed for the highlight. Antwerp Blue is used to darken the underside of the handle parts and the area behind the clear cover. When the latter is dry, the entire cover is overlaid with a Steel marker and then again on its left-facing surface. The Black marker is used again on the left-facing surface of the black part.

2

3. The front face is masked off with tape and some white pastel dust is wiped over it, grading slightly from the bottom upwards; the same is done on the top leaving the back corner brighter. The pastel is rubbed away over the clear front face to leave some bold reflections (this can also be done with the Steel marker) and removed from the handle reflection (because this is in front of the reflected light).

3

4. White crayon is used to put in the highlights along the radii, keeping them sharp on the glossy parts and slightly diffuse on the matt-black front. The shut-lines are defined with a black and a white crayon, and the blue parts are sharpened with a dark blue.

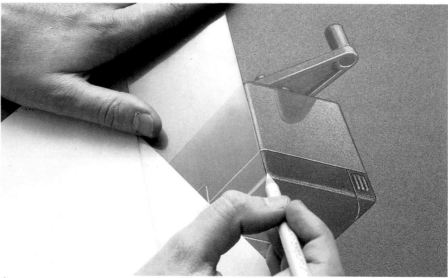

4

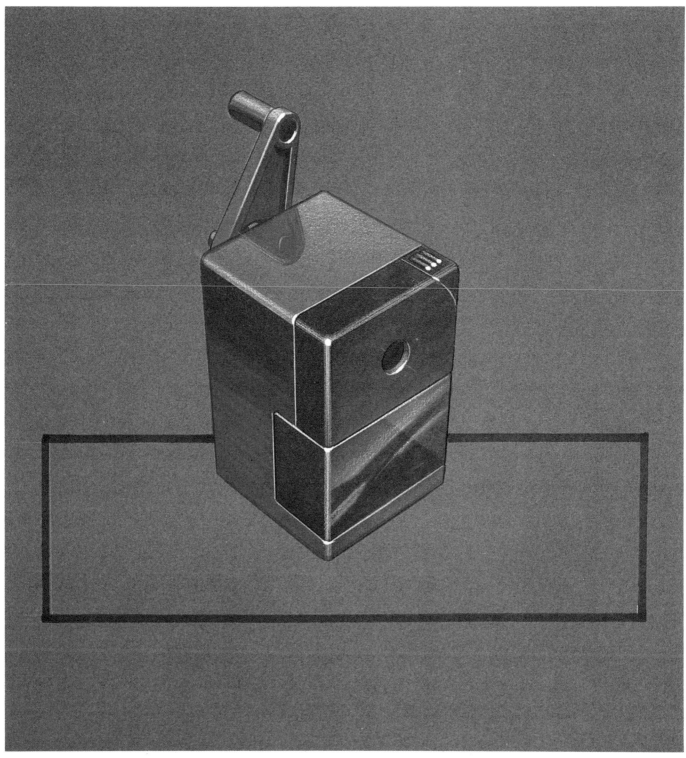

5

5. The highlights are finished off with white gouache. No attempt is made to 'ground' the drawing, rather a simple rectangle is drawn to settle it on the paper and create depth. To add to the impression of transparency, this line could have been continued behind the transparent cover.

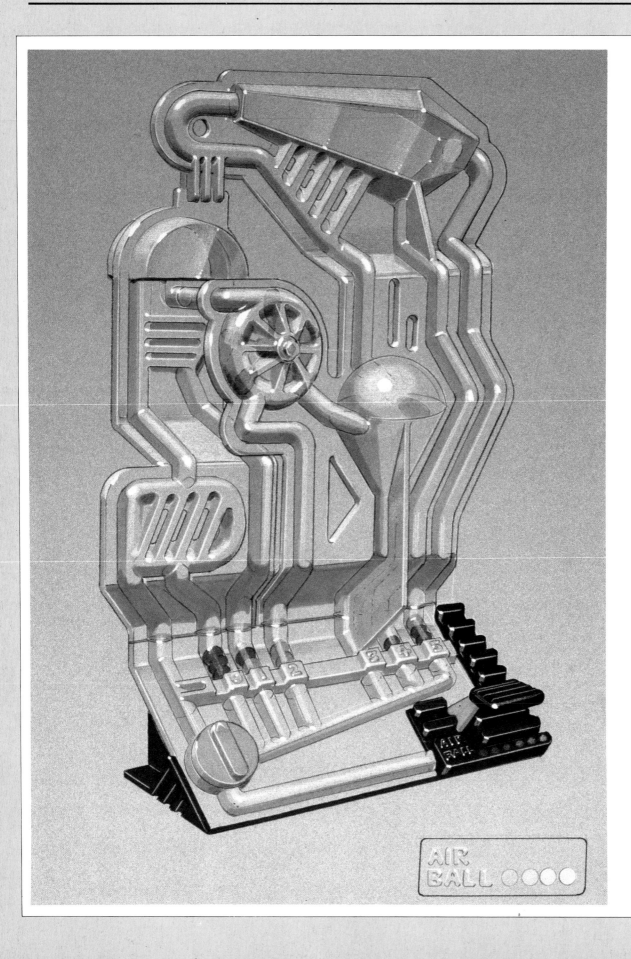

Left: Airball game
For Milton Bradley UK

Because all the major mouldings were to be transparent, this was an ideal application for coloured paper. All the tubes and shapes were lightly modelled with grey markers and a dark pencil, while the upper-facing areas were modelled using a white pencil, and highlights applied with white gouache. The drawing is not strictly correct because the black moulding runs behind the clear console moulding and should be visible behind it; for the sake of clarity this was deliberately left out.

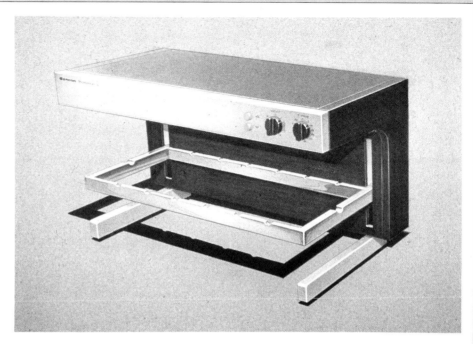

Below: Compressor
by Jim English/Ogle Design

This is a good example of making the colour of the paper the product colour and just using crayon and pastel to model the form. Markers were used for the black panel, the control area, the tyres and the jet black shadow which seats the drawing and lifts it off the background. The white pastel, adjacent to this shadow in the foreground, helps to add contrast and create the impression of 'ground'.

Above: Grill
For Canon Industries
by Peter Stevens

The grill was traced down using white chalk rubbed on the back of the underlay. The grey Ingres paper acts as the predominant colour of the unit, with marker used only for the deep shadow areas. The top edge reflection was graded with white pastel, and white and black crayons were used to tighten up the drawing and model the finer details.

Left: Circular knitting machine
For Camber International
by Jim English/Ogle Design

The concept had already been approved by the client and this was to be the final drawing before proceeding to a model. The perspective was set out by eye on layout paper and transferred to the Ingres paper with tracing-down paper. The block tones were established with Cool Grey markers but the bulk of the work was done with pastel and crayon and finally highlighted with crayon.

Above right: Sports car (the 'Maya')
by Ital Design

Ital Design in Turin are well known for their expertise in coloured paper rendering (nearly always blue). The Maya is a steel-bodied sports car designed from the ground up for eventual mass-production. This elevation of it was drawn onto blue Ingres paper. White pastel and crayon were used for the upward-facing surfaces above the horizon and darker pastel and crayon on the lower-facing surfaces. Markers were employed for the reflection in the glass and for the tyres and lower plastic mouldings.

Right: Incinerator
For Hodgkinson Bennis
by Ogle Design

This is a much larger object than the other examples, and so the view was set up with the eye-level/horizon running right through the middle, and size-impression established by fairly severe convergence towards the vanishing points. The grey Ingres paper shows through on the metallic panels on the side and front, with the latter washed over with white pastel. Note how the impression of reflective metal is enhanced by the reflection of the adjacent dark band in the top surface of the shelf. The remaining panels were coloured with marker and pastel and highlighted with white crayon and gouache.

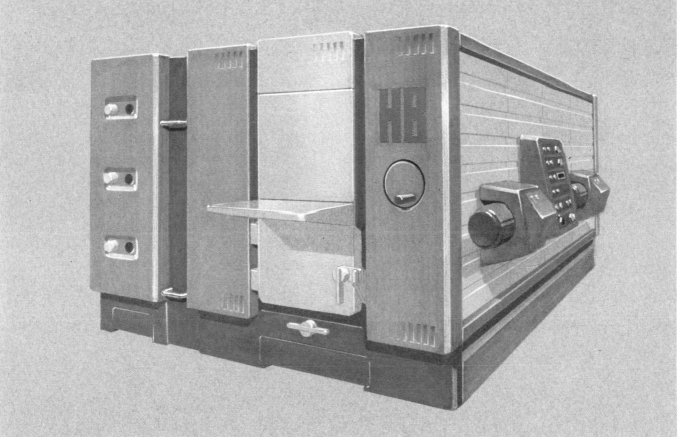

8 Automotive Rendering

There is no specific technique associated with automotive rendering, but it has a chapter to itself because it has developed so much faster than any other area of design. When it comes to drawing and rendering, the automotive designers are the pathfinders for the rest of us; techniques developed in the car studios filter downwards into industrial design studios and colleges. This is because much of the car designer's role is dedicated to styling and, wherever styling is important, so drawing and rendering skills also become important. Short of making a model, it is impossible to communicate subtle changes of form, both to yourself as you work and to others, if you cannot draw them.

Sadly, there is a fair degree of animosity between auto designers and product designers, particularly in Europe. The auto designers would argue that the product designers have lost their artistic roots and try to spread their expertise too wide, so that they become average at everything rather than good at one thing. The product designers, on the other hand, view the car designers as mere stylists, whose work is shallow and superficial, and who are unconcerned with real design problems and intellectual issues. The two views are ridiculously polarized, and both sides should have more respect for the abilities of the other. The product designers in particular should not be so quick to denigrate the work of the stylists; if they had half their styling skills, the shops would be full of products that are good looking and desirable, as well as functional.

Philosophy apart, the method of working in the auto industry is different to most other industries; in particular every aspect of presentation is more finely honed – even the loosest sketches are visually informative and graphically considered. Because the process is so 'rendering intensive' there are, inevitably, a lot of clichés. Those within the industry read right past the clichés to the design being portrayed, while those looking from outside are often bewildered by them. For example, car drawing relies heavily on visual experience, in other words, because we know that a car has four wheels and that each wheel has a tyre, there is, therefore, no need to construct and render the tyres or underside details of the car. As a result, nearly all car drawings have their tyres completely blocked-in in black, or treated in some other abstract way. Similarly, it is difficult and time-consuming to draw the interior of a car as it might be seen through the glass. To save time, therefore, many car designers lay bold reflections across the

glass. There is nothing wrong with this because they are concerned with communicating the overall *design* to themselves and to others in the studio. It is only when visually inexperienced outsiders see such drawings that they become open to misinterpretation.

Serious students of rendering will already have seen, and experienced, the many techniques used in the car industry; what follows in this chapter will serve as an introduction to a subject which, in itself, is big enough to fill a book.

Example 1:
Marker on Vellum

Sports car
by Ken Melville

Vellum is a semi-transparent tracing paper which has a coarser surface-texture than most tracing papers and, consequently, will accept crayon and pastel, which would skid across the surface of most drafting papers. It is also a great deal more absorbent than tracing papers (although nowhere near as absorbent as layout papers) and so accepts markers well. Marker ink will not, however, go right through the paper, although it is clearly visible when viewed from behind. Because both front and back surfaces are identical, both sides can be used. By markering on both sides the tonal range of a single marker can be greatly extended and some interesting effects obtained. Vellum is almost impossible to find in Britain but is widely available in the United States.

View
This is a deliberately easy view because the eyeline runs right through the middle of the car and there is very little angle to the perspective. Many automotive drawings are done with a low eyeline – either elevations or front and rear three-quarter views – because they are simpler to draw. A car is a difficult thing to draw and, if you are working through ideas, you need a view which can be repeated quickly without worrying too much about getting the perspective right. Indeed, if you are a beginner at car design, it is a good idea to start with a side-elevation with the vanishing point between the wheels. As you develop the details around the front, the side-view can be 'cheated' around so that it includes some of the front (although this is not strictly correct), and the perspective can

Example 1 : Marker on Vellum   111

be carried through to the rear wheels and roof pillars. The side-view can then be 'cheated' around to show the rear of the car in the same way.

The view shown here, however, is a true three-quarter view. It has been slightly foreshortened to bring out the dynamic qualities of the sports car. The car is also tilted to give the impression of movement as if under high G forces, with full opposite lock applied.

Stages

1. The underlay is traced through on to the vellum using a fine ballpoint pen; note how the quality of line becomes deliberately faded towards the back of the car to enhance the impression of speed, depth and form. The car is to be black and there is a horizon running along the body side. There is another reflection detail along the side glass with a broad reflection running down the steeply raked screen. Where there are no reflections in the glass, elements of the interior can be seen. The Black marker is applied in these areas, leaving the sills and lower surfaces of the tyres white because they will be picking up ground tones.

2. Next, the drawing is turned over; black applied on this side will appear as a washed-out dark grey on the other side. The Black marker is used to soften off the edges of the reflections and in the tyre highlights. The underside of the tyres is loosely masked off with the black tape and, using a tissue dampened with solvent, ink is drawn off the marker and on to the tissue and immediately reapplied to the masked-off area to obtain a graded effect. The Black is also used on the area of the screen unobscured by reflection to give the impression of looking right through the car and the rear window glass.

3. Some grey and dark-blue pastel dust are mixed together and used to grade the bonnet from the front to the back (the forward-facing panel is, of course, darker than the upward-facing surface which is reflecting sky) and also from left to right. Using pastel on the back of the drawing is a good idea if the drawing is small, or if the areas where it is to be applied are small, since it is difficult to control in these situations. As with marker on the back, it also extends the tonal range, and there is the added benefit that it remains undisturbed as the drawing is built up on the other side. Yellow ochre pastel and the same grey pastel are mixed together to create a ground tone and this is worked over the downward-facing areas such as the front bumper, wheels and sills.

1

2

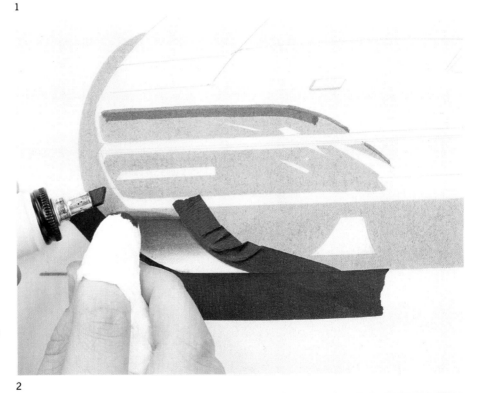

3

4.   The drawing is turned over and both pastel mixtures are worked on the front to strengthen the darker tones and increase the tonal range. A sweep is used to mask the application of pastel on the edges of the bonnet bulge, and the pastel is graded on the body side so that it gets darker as it nears the top (this gives the maximum contrast between this surface and the adjacent, but very light, bonnet surface). Using a sweep as a guide, the pastel is rubbed out on the upward-facing edges of the bumper and body panels. White gouache is mixed up and used fairly dry to highlight the facing edges of all the shut-lines, the top and bottom of the wheels, and the screen.

4

5

5.   The rectangle behind the car is traced onto a separate sheet of paper (the profile of the car did not require masking as it would be cut out later). The surface of the paper is dabbed with cleanser to begin with and then a Cadmium Red marker is streaked across, followed by a Black in the bottom left corner. A tissue dipped in cleanser is used to work the colours together and care is taken to avoid it becoming grubby. The Black is made really strong in the bottom left corner. The page is turned over and the extreme right worked at

to produce a more graded effect.
   The background is cut out and mounted using a spray adhesive. Then it is carefully cut just inside the line of the car (3-4mm) to ensure an overlap and avoid seeing the background through the drawing. The trimmed edge of the background is peeled off and the finished car mounted in place.
   The red background is reflected in the back, upward-facing edges of the car and so it appears to overlap the drawing in these areas. A white crayon or white paint is used

just to 'pull back' the edge so that the back of the car is defined.

Two observations on the finished drawing: note that the reflection of the screen in the bonnet would probably not occur in reality, but that it works well and helps to define the bonnet shape. Also note that the reflection areas are not solid black but are broken up with white streaks; this reduces their visual mass and prevents them dominating the drawing.

Example 2:
Marker

Two-seater convertible
by Peter Stevens

View
This is a two-view format which gives the client an accurate side-view (as clients have been known to scale off a rendering) and a three-quarter view to show what the car looks like as well as to provide an exciting image. Care has to be taken with the scale of the two drawings – the lower one is made slightly larger, and the higher-viewpoint drawing is logically kept on the bottom.

Because this is a finished drawing rather than a sketch, the background has to be worked out in advance, rather than considered afterwards, and in this example it has an influence on the screen reflection. Even though both views will be cut out, they are drawn together as if in their final relationship, to avoid differences of design creeping in.

The technique used is exactly the same as that described in Chapter 5 on marker rendering: the marker is put on first and then pastel and crayon used for the lighter surfaces.

1

2

Stages
1. The underlay for both views is traced off lightly in pencil.

2. The blacks are laid in using a fineliner first and then a Black marker. A biro is used very lightly to put in the shut-lines, and a Chrome Orange marker for the indicators, with a touch of Cadmium Red for the more shadowy parts. Even if a drawing is intended to be sketchy, it is always important to get the wheels right, so some colour is added to these early in the drawing – in this case a Cool Grey 3 streaked across.

3. A Cool Grey 5 is used on the lower body panels, overlaid with a Cool Grey 6 towards the rear to give some indication of plan-shape. On the upper panels a range of lighter Cool Greys are streaked boldly across, getting lighter in the middle of the side-view, and to the rear in the lower view. A white pencil is used to start to define the shapes in the sills and the shut-lines, and a combination of sky-blue and bluey-grey pencils is used on the edges where the body side rolls over onto the upper-facing surfaces. With a silver car, or any metallic finish, you tend to get a darkening (sometimes called a darklight or lowlight) on edges and curves where you would normally expect to see a bright highlight; this is because of all the metal particles suspended in the clear lacquer.

3

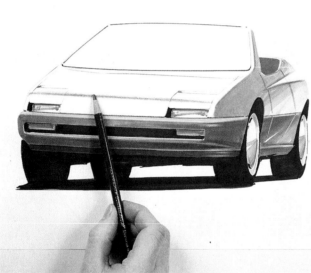

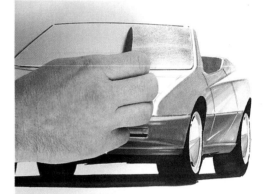

4

5

4. To get a smooth tonal graduation with the coloured pencil the paper is moved around until the most comfortable position is found for the hand, and the pencil is laid down almost parallel with the paper. The headlights are treated first of all as if there was no front glass, only a reflector. The upper surface of the reflector is facing downwards and so will be reflecting ground tones, which are put in loosely with a Warm Grey 4 marker.

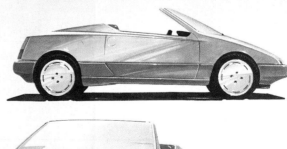

5. With care, and depending on the paper being used, it is quite possible to use pastel direct from the stick, softening it afterwards with tissue or cotton wool. This must be done very gently to avoid scratching the paper and leaving a pastel line that refuses to blend; if this is too difficult, the scrape/dust/cotton-wool method can be used as in other examples. Where the pastel runs off onto

6

7

white paper (for example, on the bonnet and screen of the upper view), the blue-grey pencil is used to tighten and darken the edge.

6. Behind the screen of the lower view, some of the interior would be visible through the reflection, so this is suggested on the right side with a red pencil. On the other side, where something dark is being reflected, the dashboard would be clearly visible through the glass; this is put in with a grey marker. All the details, such as the wheel nuts, are finished off and some rudimentary tread details are put in on the tyres with a white pencil. The shut-lines and edges are crispened up with a black biro, lightly applied, and the highlights are finished off with gouache in the usual way. The photo shows the drawing finished and ready for mounting with the background shape cut out of tracing

paper and lined up underneath. Note how the tracing paper shows very clearly underneath the drawings. The background should always be cut away behind the images to avoid this happening.

7. An appropriate black-and-white newspaper photo is chosen and lined up with the tracing paper background until the overall composition is satisfactory. A grey car paint is sprayed very lightly on to the tracing paper and allowed to dry. The front (yes, the front) of the photo is sprayed very lightly with Spray Mount and stuck face-down onto a separate sheet of tracing paper; this has the effect of toning down the photo so that it looks slightly hazy. The grey background below is stuck down lightly and the photo with the finished drawings tacked down in the required position on top. The two views are carefully

trimmed out using a sharp scalpel, which must also cut through the background at the same time, so that the two parts will fit snugly together like a jigsaw when mounted down to the final board. The screen on the lower view is cut away completely to define the end of the reflection and to reveal the background behind; the top edge of the screen is put back with coloured pencils.

Example 3:
Coloured Pencil

Three-door family car
by Tony Catignani

Coloured pencil is a very fast and 'loose' technique that is ideal for working up ideas; the design can evolve as you draw because the medium is flexible and allows alterations and changes of detail to be made.

View
This is a straight side-view based on a Ford Fiesta 'package' (the operational and dimensional specifications).

Stages
1. The design is worked up freehand off an underlay, using crayons. Note the exaggerated perspective which helps to make the straight elevation more interesting. The wheels are tied down tight using a circle guide as they form the basis for the whole drawing.

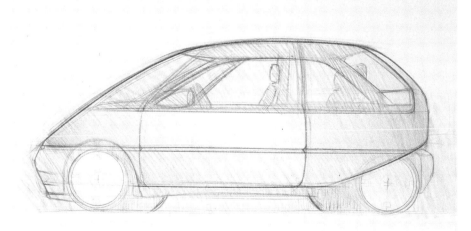

1

2. Working quickly and fluidly, the drawing is built up using bold passes of the pencil laid flat to the paper surface. The centre section is kept white to form a 'hotspot' and black added over the red at either end to make them darker. Tension is brought into the line-work by tightening up with a sweep rather than a straight-edge. As the drawing develops, and it becomes clear which tones to lay where, a darker pencil is used. The lower body panels are going to be a darker 'stone chip' protective plastic and so these are put in with a dark blue crayon.

3. A Cool Grey 3 is used across the glass (there are no body-colour pillars) and across the wheels and background, and then on the body panels so that the marker bleeds the crayon effectively. At this stage the finished drawing is beginning to emerge, and a closer look can therefore be taken at the tonal balance, composition and so on. A Cool Grey 6 is used to darken under the car, and around the wheels a bit more. The crayon work is blended with a tissue and more red crayon is overlaid across the body side to increase contrast. To make the glass look tinted, a yellow pencil is applied right across it.

4. The wheel and door-handle details are put in and the door shut-lines hardened up with a dark crayon. A bright red crayon is used where the window meets the body, fading it towards the middle, to create a little ledge of metal. White and black crayons are used together to simulate the flush glass appearance, and the black pencil alone to

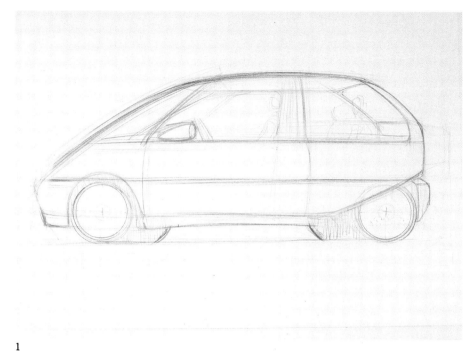

2

darken the underside of the wheel arches and create more contrast. It is used again on the lower edge of the body panels where they turn under. White gouache is mixed up and a drop of washing-up liquid added to break down the surface tension, and so prevent the paint blobbing. The paint is used fairly 'washed out' along the top of the body mouldings and the tops of the tyres, but a stronger mix is used for the highlights. The highlights in the corners of the glass are like signatures which tell us that the glass is dropped in flush.

5. After the drawing was mounted down, it was decided that more contrast was needed around the wheels and under the car, and so this was put in with a Cool Grey 7. Finally, an orange crayon was worked across the middle of the body panel to liven up this area.

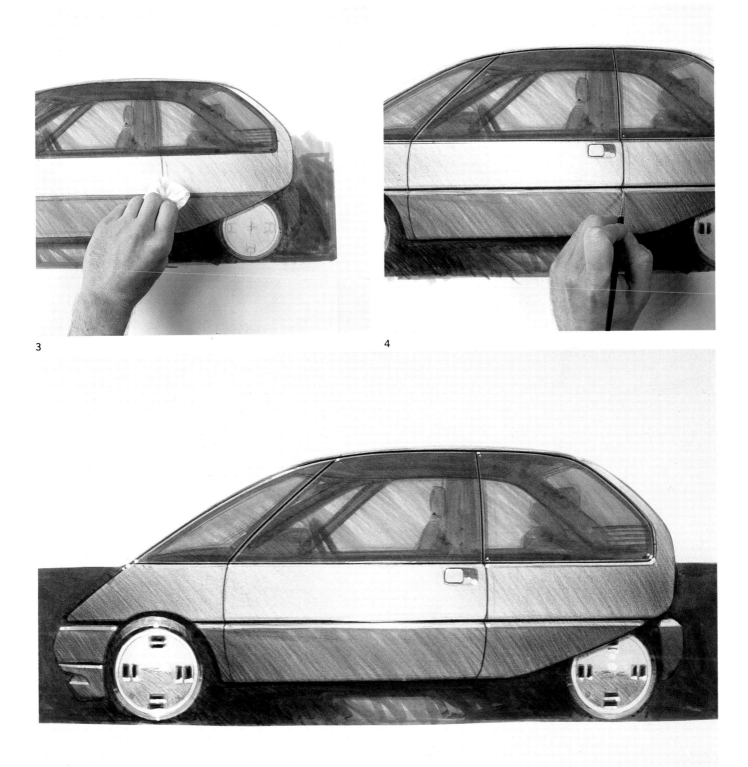

3

4

5

Example 4:
Full-Size Tape Drawing

Two-seater city commuter
by Ken Melville and Gunvant Mistry

In the automotive business the full-size tape drawing is the intermediate step between renderings and fifth-scale package drawings on the one hand and the clay model on the other. A drawing is a lot quicker and cheaper than a model and allows several alternatives to be reviewed so that one can be progressed to a model. Because the drawing is based on a true 'package' (the dimensional, operational and ergonomic details of a car) there is no possibility of 'cheating' and, like

the generator on page 73, the drawing can give a true indication of size and proportion. Surprisingly, what looks good at fifth scale needs a fair amount of modification to make it work well at full size.

To those who have never done a tape drawing it may seem a mystery why anyone would wish to use tape to actually draw with, but, as soon as you try, you understand why: you can make extensive changes to the shape without constantly erasing and redrawing, you can easily lay down a curved line and, if you are unhappy with the rate of curve, peel it off and try again.

When using tape in this way, the trick is to anchor one end and keep it in tension with one hand as you use the other to guide the

tape around the curve. To help get a smooth curve it is sometimes useful to create the curve using a thicker tape (which bends more consistently) and then lay the final, thinner tape butted up to it; the first tape can then be removed.

Package

A two-seater city commuter, wheelbase 2m (6ft 6in), height 1.33m (4ft 3in), overall length 3m (10ft). Called Ciao City. Based on a BMW 1000cc motorbike engine: flat four transverse, approx 90bhp.

Preparation

Choosing a site. Make sure that you have plenty of room to operate, with sufficient blank space around the drawing to avoid cramping the image. You will need to step back from the drawing to give yourself an 'at a glance' appraisal of how it is progressing; this needs at least twenty feet if you are to take in the whole drawing without having to turn your head.

Paper. Vellum, Crystalline, or one of the Polyester films work best, with the first two absorbing much more ink than the Polyester, which is virtually impervious. Polyester film is expensive, but dimensionally stable and very tough. The drawing is taken off the wall and replaced several times as well as being subject to constant abuse as tape is ripped off and masks are cut, so a tough paper is essential.

Stretching the film. It is important to stretch the film very firmly because as you lay down tape you put it in tension, and if the film were not stretched it would cockle. Stretching is done at the mid-point of each end, and then at the mid-point of the two longer sides; these are all held with 2in tape and then a couple of staples put through both tape and film for final stability. The corners are then stretched in the same way so that the whole is like a drum skin.

Transferring the design. To help in the transfer from the fifth-scale package to full size, a 100mm grid is drawn first (film which is pre-printed with a grid is available) so that the view can be accurately scaled. Be prepared for a dramatic change to the design because your perception of proportion and detail will change once the design is roughed up in full size. Be sure to step back as far as possible to give yourself a broad view of the whole drawing.

Planning. Work out exactly in advance what you intend to do; you will find there is very little room for manoeuvre once you begin. Render up the drawing at fifth scale in almost the same way as you intend for the full size, and practise the detail areas if you have any doubts. This example is to be a black car with a horizon along the body side. Running above the horizon, and in the glass, is a dark band;

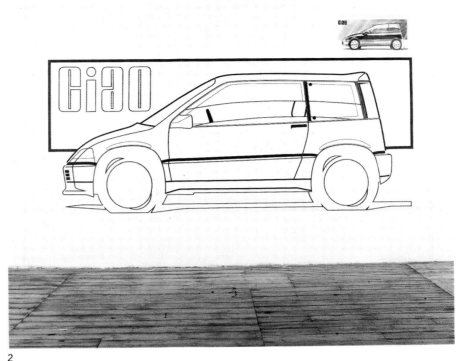

2

this is sometimes called a 'double horizon' but could equally represent something long and dark in the foreground which is therefore obscuring the horizon behind it. Because this band is dark, we can see right through the glass to the background behind the car. On the other hand, where the glass is reflecting light, we see nothing through the glass. The background is there to emphasize this and therefore help the glass to look transparent, as well as throwing the drawing forward and providing warmth to contrast with the cool colouring of the car. Its sharp, rectilinear form also helps frame the car and 'settle' it graphically so that the composition works well.

Stages

1. *Tape drawing (1).* The design, and the planning for the rendering, are well worked out at fifth scale first before even contemplating moving up to full size. Once happy, and reasonably confident, a grid is drawn onto the fifth-scale view – either on a clear overlay, or simply drawn over the top, so that this can be scaled up to full size. The grid is then drawn five times up on the stretched film and the design transferred to it using a pencil to rough in the lines, followed by tape. The photograph shows this stage in preparation. Note the three reference drawings pinned above.

During this stage the film is subject to a fair amount of abuse as tape is constantly applied, removed and reapplied. This can

leave traces of adhesive and grease (from hands) on the surface, so, to be safe, it is best to do the final drawing on a fresh sheet. Once the design looks right, a new sheet is stretched over the top and it is re-drawn taking into consideration this time round where the first application of colour will be.

2. *Tape drawing (2).* This shows the drawing on a fresh sheet of film and ready for the first application of colour. Note how the window horizon is clearly defined but that along the body side has yet to be taped, although the wheel-arch eyebrows are in.

Next, those parts are identified which will eventually be white in the finished drawing and can therefore be masked off and forgotten about until near the end. This does not include broad white areas which need light airbrushing, only those small areas which are well defined such as the door shut-lines, upper surface of the body panels and bumpers, and the split line between the two lower panels.

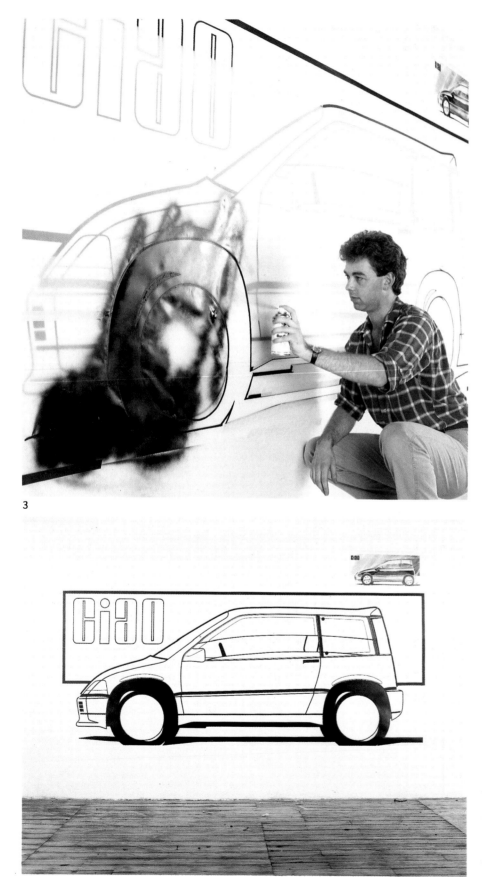

3

4

3. *Black (1).* All the black areas are carefully masked out using lightweight tracing paper which is overlaid across the drawing as far as the spray is likely to carry, and then a bit more. The black areas are then trimmed out with a scalpel along the middle of the underlying black tape. The tracing paper is held down along its entire length with more black tape to prevent the paint getting under the mask. Very small areas (such as the tyre highlights) should not be masked out with paper; only tape is used to cover over the underlying white. The entire area is then sprayed with paint (in this case a Marabu acrylic) and allowed to dry.

4. *Black (2).* This photo shows the masking paper and tape removed.

5. *Working on the reverse side (1).* The body-side reflection is put in using a thin tape so that when the drawing is turned around it is easier to see where to mask. The staples are lifted out and the sheet reversed, re-stretching it as before. The body-side reflection (including eyebrows and reflection detail at the central 'hotspot') and the shadow under the mirror are masked out. Also masked at this stage are the details within the car that are not obscured by reflected light: the headrest and the insides of the A and C (front and back) pillars. Finally, the small rear-light detail on the roof will reflect the same information as the body side, so this should also be masked. All these areas can now be sprayed with the same flat black as the wheels.

The lower parts of the two indicators are masked out and some orange Flo-master ink wiped in, and some sky blue is also applied to the headlights (perhaps a little too strongly).

The highlights on the tyres are sprayed black from behind so that they appear grey on the front side. The small amount of interior (roof lining etc) is also done at this stage using black Flo-master and cleanser so that it fades out somewhat from the top to the bottom, to match the sky graduation that will be applied later on the front.

The base grey of the body panels which have been masked out is done with Flo-master on the back so that the real modelling can be done from the front. When wiping on colour in this way it is a good idea to use straight cleanser to start with to get the ink flowing. The grey is wiped to the same central 'hotspot' as the body-side reflection, and once this masking is removed, this hotspot is sprayed with white Marabu acrylic to reinforce the impression of a sunset/sunrise coming over the horizon.

5

6. *Working on the reverse side (2).* All the tapes which were aiding masking on the back are removed so that there is no black edge to the shapes once the film is turned around again.

7

7 and 8. *Background (1).* The paper is re-stretched and the background masked out including the lettering. Where there is no light being reflected in the glass, you will see right through to the background, so these areas are also masked out. The screen and rear window have been masked to allow the background to be reflected as this will suggest the amount of curve in the plan-shape; a tape has been left so that after colouring a white line will be revealed. Finally, the rest of the drawing is thoroughly covered over with tracing paper. Photo 7 shows the drawing during the first application of colour which is wiped on boldly from the top left using a 30cm (12in) giant marker. To prevent the colours becoming muddy four separate swabs are used, one for each colour, and one for cleanser only. Photo 8 illustrates how important it is to cover everything!

8

9. *Background (2).* This photo shows the drawing at exactly the same stage but with the masking removed.

9

10. *Body-side reflection (1).* The areas above the horizon and the wheel-arch eyebrows are airbrushed next using inks. All but these areas are carefully masked out with tape and tracing paper ready for spraying. The eyebrows are sprayed to a central highlight so that they nearly blend with the underlying black at either end. Blue ink is loaded into the brush immediately after the black so that the area nearest the highlight has a bluish tinge; once complete and dried, they are covered over for protection from overspray.

10

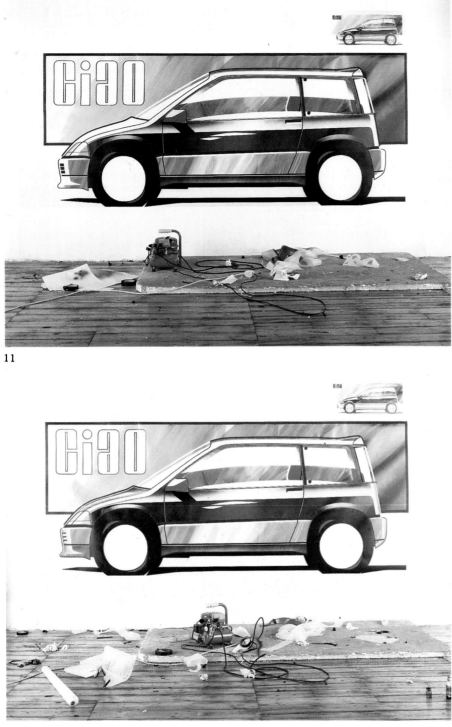

11

12

11. *Body-side reflection (2)*. The area above the horizon needs careful planning and we elected to leave a broad white highlight along the radius, up the A pillar and right along the roofline, so this is masked out first. The rest of the area needs to be toned towards the same central hotspot as on the lower panels. Before beginning to spray, some tape is laid along the top of the body panels (just below the glass); this will be removed after the first few passes with the airbrush to leave a slightly lighter part where the bodywork curves into the window and is therefore picking up reflected sky tones. As with the other black areas this is oversprayed with blue loaded into the brush on top of the black.

Looking at the drawing at this stage, we decided to darken further the area below the horizon, as this required some modelling towards the centre to match the adjacent areas and emphasize the plan-shape. This was therefore masked up and sprayed until the tone of the black in the darkest areas matched the newly sprayed areas above the horizon.

12. *Bonnet and rear*. The bonnet is masked up next so that the reflection used on the screen is carried through and down to the headlights, but deflected by the kick-up in front of the screen. The rear of the car is also sprayed during this phase as the airbrush is ready-loaded with black. The front and rear bumpers are sprayed next, starting with the darker, under-facing surfaces and then the front- and rear-facing edges. Finally during this stage, the hotspot is masked out and Marabu white acrylic paint sprayed to create maximum brightness.

13. *Window reflection*. The area above the reflection in the glass needs airbrushing with sky blue and so this is masked out and very lightly sprayed to match the grading done earlier on the insides of the pillars. As with the body-side reflection, white acrylic is used to brighten the hotspot (being very careful to avoid runs).

14. *Wheels and details*. The wheels are done completely separately on layout paper (Frisk CS10 paper would have been better but was not available). The radial slots are divided up into easy areas and sprayed in sequence from dark to light so that masking does not need replacing. They are then given a complete graded tone from top to bottom to suggest a gentle curve. The lettering is cut from red Chromolux (high gloss) and stuck down before finally sticking the whole wheel down to foamcore and trimming out. The finished wheels are attached, using sticky pads.

All of the small details, such as the window 'roundels' are put in using local masks (also the top of the mirror which was forgotten earlier). The headlight needed quite a lot of airbrushing to overlay the slightly greeny-blue colour that was erroneously applied on the back of the drawing, and restore it to a more sky-blue colour. White tape is used along the shut-lines and on the top edge of the lower body panel (in the middle) to suggest a highlight. The front fender-line looked too harsh and was re-masked so that it could be further softened off. Finally, the rear light was sprayed using local masks.

13

14

15

15. *The finished drawing.* The drawing is removed from the wall so that the entire presentation area can be cleaned up, and is then re-stretched with the wheels firmly on the ground.

Interiors

Within the car industry there are often two completely separate styling/design departments: one for exteriors and one for interiors. While there is obviously a lot of cooperation between the two, there is also a fair degree of rivalry. The exteriors department is looked on by many as the glamour side of the business, and most aspiring car designers want to start their careers there. In reality, the challenge of designing interiors is a great deal more demanding and, in many ways, far more like an industrial design project. A car exterior is basically a single component with five or six extra parts, such as bumpers and mirrors, whereas an interior is made up of perhaps seventy components, each of which has to be considered and designed in detail, and yet fit and work together as a complete interior.

Generally speaking, more visuals are done for exteriors than interiors, because an exterior is more about shape, form and other emotive issues, whereas an interior has to address the problems of packaging, detailing and the coordination of disparate parts. As a result, the types of drawing differ slightly, with interior renderings tending to be more like product drawings than the more stylized exteriors.

An exterior visual takes a lot for granted; it relies on the viewer to fill in much of the detail from visual experience. A line here, a reflection there, and an image can be constructed from almost nothing. As a result it is easy to cheat in order to show off the best features of the design: wheels are grossly enlarged, ride heights lowered, tyres widened and so on. With an interior visual, on the other hand, it is very hard to cheat – you can't leave out details because the eye simply doesn't fill in the gaps. Even parts not strictly relevant to the design need to be included for completeness, otherwise the drawing is like a face without a nose. Depite this, there is still a lot of the car designer's art in the design of an interior, the basic issues are the same, and the emotive and styling input is as strong. Few designers of interiors will therefore settle for a wooden portrayal of their design on paper, and will strive to produce a dynamic and stylish drawing that is every bit the match of that produced by designers of exteriors.

Example 5:
Car Interior
by Ian Callum

View
In the exteriors department, drawing styles come and go like fashions, but in interiors, changes of technique tend to evolve more slowly (partly because there is less actual drawing done). For this reason it is easy to produce a boring view of an interior, particularly if the same basic view is constantly recycled as an underlay. The view chosen for this demonstration is therefore slightly unreal to make the drawing look more dynamic; it is also economical because the steering wheel is an almost perfect circle, and therefore quick and easy to draw. Like the wheels in an exterior sketch, it is very important to get the steering wheel right; because it *is* such a focal point it pays dividends to put effort into it. Graphically, the drawing is comfortable on the page, because the lines which bleed off top right and bottom left will run parallel to the edge of the paper.

Planning
Lighting is quite a problem on a drawing like this because, in reality, the light would come in fairly evenly from all sides through the windows. Because of the slope of the screen there would be a fair amount of top light on the dash panel at the front, which would fade into deep shadow near the floor. The result of such all-round lighting is to soften off shapes and reduce contrast, neither of which is really desirable if you are trying to describe detail. For this reason it is better to adopt a convention that makes the task easier, such as imagining that the roof has been removed and that the light source is more consistent. There will still, however, be a lot of light being reflected off the top of the dash, and the lower areas will have to be rendered a great deal darker to give depth to the view. To add to this illusion of depth, the drawing will have to gradually lose detail as it gets lower; equally, it will have to blend out and lose intensity as it reaches the edges, with the focus on the wheel and dashboard, rather than the whole interior. With a drawing such as this, you are illustrating a design/theme rather than definitive recommendations for colour and trim which will come later. Liberties can therefore be taken with the colour (such as grossly overlighting the dash) to focus attention and prevent the drawing becoming flat and dull.

Stages
1. This is the first sketch, drawn with a coloured crayon, to block in the design. The view is established and the details of the design laid in. Once happy with the layout and content, the underlay is traced off (again with a grey Prismacolor crayon), truing and tightening at the same time.

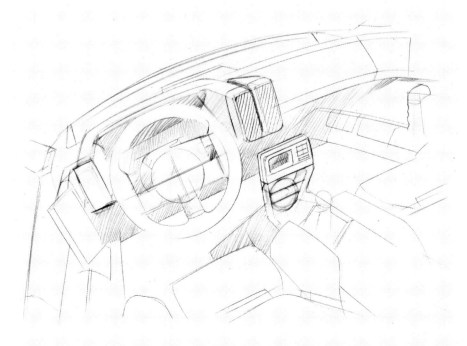

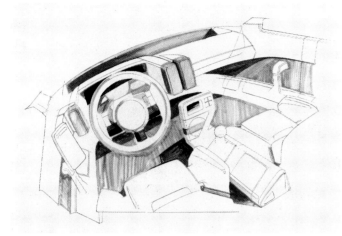

2

3

2. Cool Greys 2-5 are used on the doors, floor, binnacle (instrumentation) and cowl (top of dash) areas to establish basic tones; some of these will remain grey, while others will be overlaid with colour (both marker and pencil) later on. A fairly loose and sketchy application of marker is aimed at to keep the drawing lively, so it doesn't matter about the

patterns left by the marker strokes; however, this effect should be contained within each area rather than letting the marker run over key-lines.

3. A Lipstick Natural marker is used over the grey in the lower coloured parts and seats, adding Pale Rose and Flesh colours in the darker areas.

4. A red coloured crayon is used over the marker, particularly in the shadow areas, to take the greyness out of the colour, and the same crayon is used to put in detail lines such as shut-lines and edges. The whole of the dash and the upper parts of the door panels are treated with pink pastel and rubbed away to produce highlights on the radii and upper-facing surfaces; the sides of the seats are also treated in the same way. At this stage the top surface of the binnacle can be dusted with a light-grey pastel (because there is more colour, such as sky tones, being reflected through the screen), and all the lighter surfaces of the other grey parts (such as the centre console, wheel, control pod) with a plain light grey pastel. As an alternative to drawing everything at once, it is often a good idea to draw the binnacle, instruments and controls separately and stick them down later on. This allows you to treat both parts differently with no need to worry about pastel spreading from one to the other; it also means that you can add considerable detail to the binnacle, and therefore focus attention on this key area.

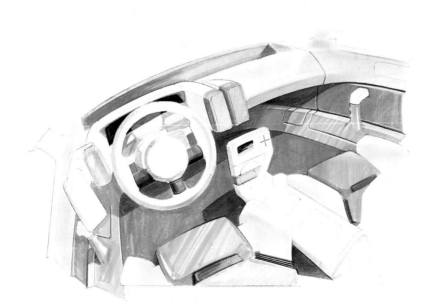

4

5. To help create the illusion of a textured material it can be very effective to slip a real texture beneath the drawing and work over it with a coloured crayon. For carpets and rough textures, sand paper or emery paper can be used; for specific textures such as leather it is best to look around for moulded 'look-alikes' that are hard enough to transfer through rubbing and are the right scale. In this example, a white pencil is used to reveal the texture of the moulded rubber bumps beneath the paper. By moving the texture very slightly and using a dark red pencil, the darker side of the bumps can be simulated to give a more three-dimensional effect.

6. The finished drawing. All the shut-lines and general line-work have been tidied up with coloured crayons, and some reflected colour used in the binnacle. The gear-lever gaiter has been put in with a Cool Grey 3 and detail modelling of the wheel and controls completed. Highlights are laid in with white gouache and the drawing trimmed out and remounted to white card.

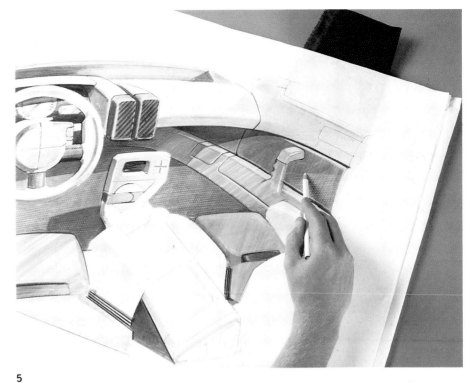

5

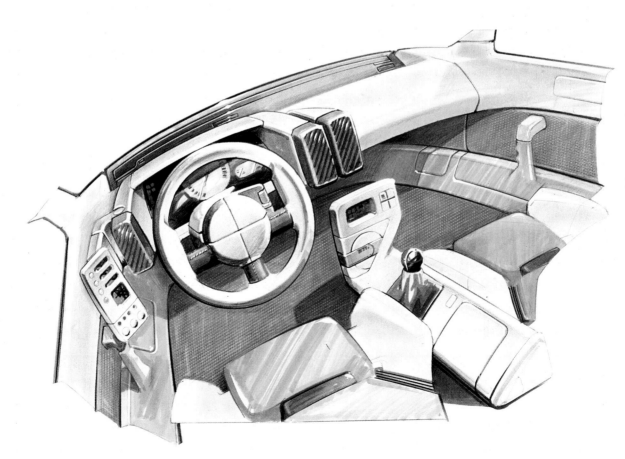

6

Above and above right: Four-door saloon
by Yukio Iijima/ Ford Motor Co.

This is an ambitious drawing incorporating, as it does, a complete reflected image, which almost means that it has to be drawn twice. This is fairly straightforward to construct provided, as here, the eyeline is sufficiently low down to allow the underlay to be simply flipped over and traced through. The combination of hard, crisp reflections, such as in the side glass and body side, combine easily with the softened edge of the pastel to produce a good example of the soft forms associated with contemporary cars. The annotated drawing shows the importance of working on both sides of the paper as the designer builds up the tonal depth needed to describe this type of complex form.

Right: Two-door saloon
by Pinky Lai/Ford Motor Co.

This is a good example of a black car. Note the rich blue sky tones reflected in all the upward-facing surfaces, and the bold window reflection revealing the interior.

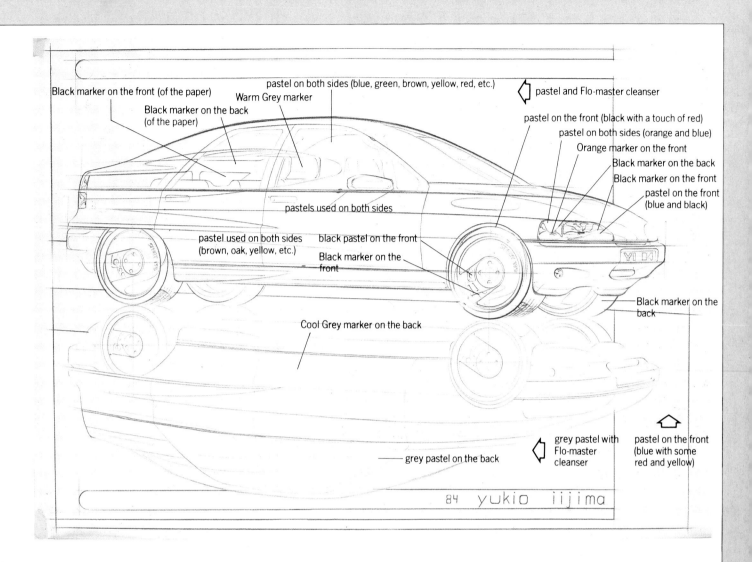

Black marker on the front (of the paper)

Black marker on the back (of the paper)

Warm Grey marker

pastel on both sides (blue, green, brown, yellow, red, etc.)

pastel and Flo-master cleanser

pastel on the front (black with a touch of red)

pastel on both sides (orange and blue)

Orange marker on the front

Black marker on the back

Black marker on the front

pastel on the front (blue and black)

pastels used on both sides

pastel used on both sides (brown, oak, yellow, etc.)

black pastel on the front

Black marker on the front

Black marker on the back

Cool Grey marker on the back

grey pastel with Flo-master cleanser

pastel on the front (blue with some red and yellow)

grey pastel on the back

84 yukio iijima

Above: Four-wheel-drive ATV (All Terrain Vehicle)
by Peter Hutchinson

Peter Hutchinson entered this sketch for a competition while still a student. It is a bold, low-level view with a loose, sketchy feel. Rendering a white car on a white background is difficult enough and this drawing is both informative and exciting – a good example of 'minimum marks for maximum information'.

Above right: Truck interior
For Leyland
by Marcus Hotblack/Ogle Design

This proposal for a truck interior was done from a high three-quarter view from somewhere outside the cab, as if there were no roof or back panel. It is, therefore, a strictly imaginary view which was chosen to afford the maximum information about the contouring of the panel-work, seat and floating binnacle. (The steering wheel, itself the subject of a separate rendering, was not shown in order to avoid obscuring the design of the panelling and binnacle behind). The drawing was done with Magic Markers and highlighted and toned with coloured pencils. Bright yellow Flo-master was used in the background to 'punch up' the drawing.

Right: Advanced transportation concept study for the year 2000 and beyond
by Pinky Lai

These are magnetic levitation vehicles which run on special roads. The vehicles themselves are black, and visual impact is achieved by contrasting the black with the sharp, even shocking colours of the background which are also reflected in the vehicles. The background can be put in with either marker or pastel and Flo-master cleanser or Flo-master ink, or a combination of all three. The transitional areas were done with a black Prismacolor pencil fading out from the black marker towards the contrasting colour. Finally, all the shut-lines and corners were highlighted to 'punch up' the drawing.

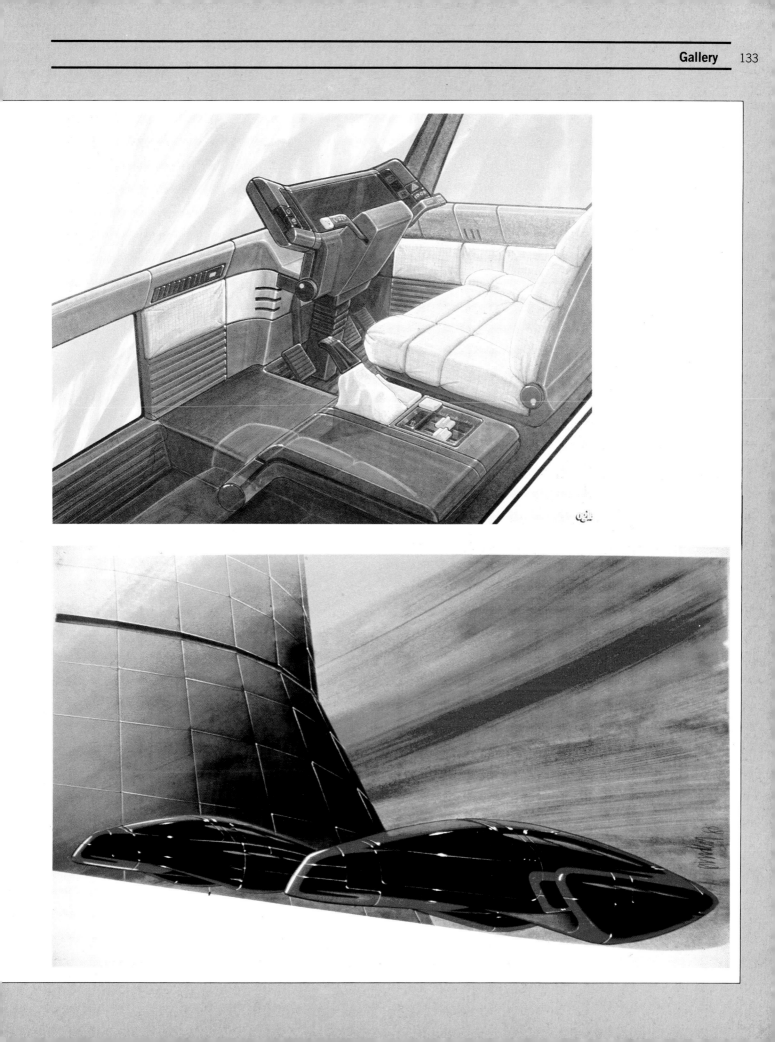

Action Vehicle
by Julian Thompson

*This dynamic and exciting drawing illustrates
the spirit of the product as much as the overall
look of the car. The design was drawn out
onto Crystalline paper using a ballpoint pen
and then marker was laid on both sides of the
paper, particularly where more colour
intensity was required, such as on the yellow
bumper. Towards the back of the window
reflection and the car itself, the marker was
washed out using Flo-master cleanser. Pastel
has also been used on both sides, with local
spraying of fixative between applications to
prevent smudging and to allow different
colours to be overlaid and abutted. The
blending of tyres and white paper makes the
car look lower and was done loosely to
suggest long grass. The front bumper looks
like textured plastic, and this was achieved by
using a Cool Grey 8 on the back and then
overspraying with fixative to give a slightly
dotty finish.*

Roadrunner truck exterior
For Leyland
by Charles Harvey/Ogle Design

This rendering was done after the overall concept was approved, in order to validate proposals for detail parts, such as grills, spoilers, etc., which visually define different models in the range. The viewpoint is at eye-level, and there is plenty of perspective convergence across the front, to give a dramatic effect to the view, while still telling us a lot about the design of both the front and side.

The drawing was done with Magic Marker, with some pastel for the softer tonal transitions. Note how the windscreen reflection is carried through to the bodywork, and that the red background is clearly visible through the windscreen, but not through the area of bright reflected light, where only an impression of the interior remains. The black rectangle running behind the truck also combines with the red background to help throw the image forward off the paper.

Three drawings of the ECO 2000 concept car from the Peugeot SA group
by Scott Yu

Above: *This drawing has two rear views, both derived from the same underlay, but with the one on the right showing the rear door open to reveal its transparent construction and elements of the interior. The dark background has been used to give contrast to the almost white upper-halves of the car; note also the low viewpoint to give an almost horizontal waistline with very little convergence to the vanishing points.*

Above right: *Most interiors tend, through necessity, to be fairly tightly drawn and defined, but this one is loose and relaxed, underlining the conceptual nature of the design. The view focuses on the binnacle and its surround, omitting the steering wheel, and was drawn with a fairly severe perspective. The inset shows the graphic readout display. The broad marker strokes below the binnacle give the impression of depth and provide contrast, without the need to draw in all the floor and pedal details.*

Right: *This is a good example of a rendered side-view done with markers and pastel; it has a nice sketchy feel and is well balanced by the washed out truck in the background.*

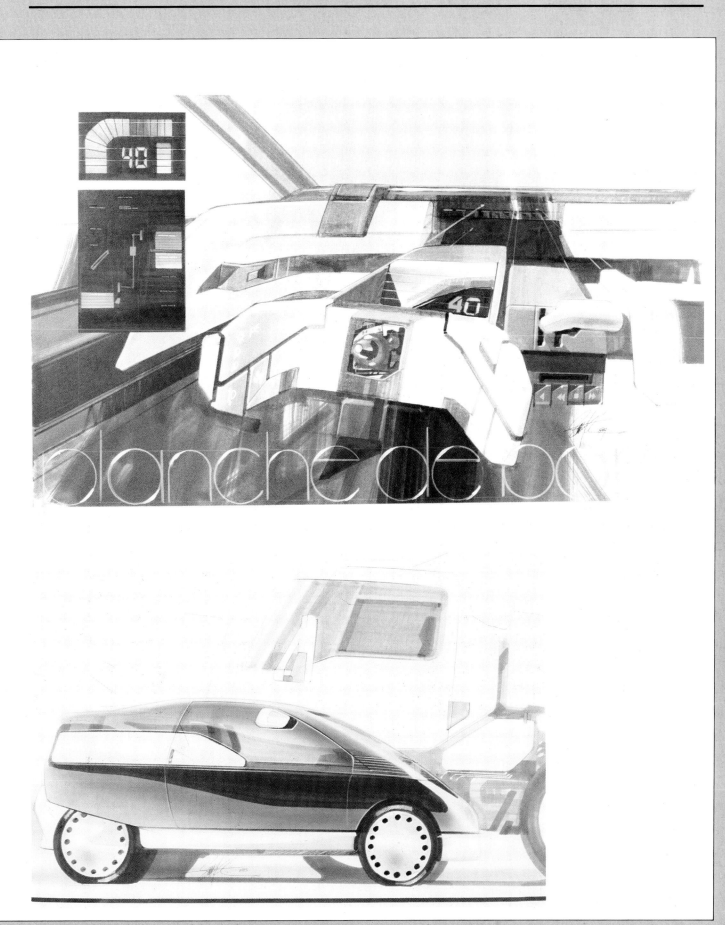

9 Special Finishes

There will be many occasions when you are faced with the problem of colouring up a specific material which you have never tackled before. There is no substitute for getting hold of the actual material or finding some reference for it. Experiment with this in different lighting, looking closely at the finish to see how reflective it is and how much the colour of the reflected image is changed by the material. If it is a complex surface (such as fabric), try looking through half-closed eyes to filter out unnecessary detail.

Wood

Furniture designers, who need to render different types of wood every day, will probably be fairly accomplished with water-colour or gouache because, for detailed work, traditional media are hard to beat. In some applications it is possible to render wood quite effectively using a streaked marker base, overlaid with coloured pencil to suggest the grain. The method shown below is ideal for large areas of timber such as floors and table tops, or elevational views of timber cabinets.

1

Transparent Materials

When drawing clear materials it helps to think of the background first, as if you were seeing straight through the product. What you actually draw is the background, modified by the presence of the product in front of it. Doing it this way round ensures that whatever you draw will look transparent, and for this reason, working on coloured papers and darker backgrounds is a good way of immediately getting the right effect. Extrapolating from this, it is easy to see that tinted glass, or plastic, has the same effect as holding a coloured filter over the background, which will therefore be rendered in tones of whatever colour tint is being used.

Mahogany table

1. Draw out the table on to a sheet of layout paper. Taking a separate sheet, scrape a selection of pastel colours into small piles; select only those colours appropriate to the chosen timber, in this case, mahogany. Using cotton-wool balls lightly soaked in cleanser and then dipped into the pastel, streak boldly across the paper to create a 'field' of woodgrain. It is a good idea to position the underlay underneath as you do this, to help keep the grain in line of perspective.

2. Trim the table top out of the sheet and use a Light Mahogany and a Mahogany marker on the two side edges. Treat the legs in the same way, but on a separate piece of paper, and stick the top down on it, using Spray Mount. Spray over with fixative and allow to dry. Then pastel over the back corner with white and rub away to produce lowlights. The fixative ensures that the mahogany pastel remains while the white is rubbed away. An alternative way of doing this is simply to rub away the table top pastel to produce highlights as opposed to lowlights.

2

Nearly all completely transparent materials have a very high gloss finish; they have to in order to allow light to pass through undisturbed. (If the finish was matt, then the light would be scattered in all directions and would distort, cloud or modify whatever is behind.) The high-gloss finish means that there are likely to be sharp, high-contrast reflections that will also obscure the background details behind. With clear materials the angle of incidence to the light is important: if you are looking through a sheet of glass, and at right angles to it, you will see right through it very clearly. If, however, you turn it gradually, so that the angle of incidence increases or decreases, you will see less and less through the glass, and more and more reflections. This is why heavily lacquered paintwork looks so shiny: near to the perpendicular you see through the lacquer to the underlying paint, but at acute angles you see crisp reflections. Where these are light, you see little of the paint, and where they are dark, you see more of the underlying colour. As with all glossy surfaces, the reflections in transparent materials are sharp and well defined, and, if anything, even more contrasty than their opaque equivalents.

The following quick sketch shows how to draw smoked acrylic.

Disc storage box

1. Draw the box freehand as if it had no acrylic cover but don't put in too much detail under the cover, just broad outlines and shapes. Leave the intended highlight areas (i.e. window reflections and radii) uncoloured for the moment. Render up the base of the box with markers.

2. Treat all the acrylic in the same way as you would opaque plastic and colour it in with a Steel marker, working over the white highlight areas (which I had originally intended to leave white). Allow to dry and work over it again, darkening near the radii highlights. (Sometimes, if you have several overlapping layers of smoked acrylic, remember to count the number of layers between you and the background because each additional layer will make that area one tone darker.) Talc over, and add a little white pastel to the front angled face and to the top. Use the Steel marker to remove this combination locally from these surfaces to produce the streaked reflection. Use a Cool Grey marker on the front vertical plane to darken it with respect to the sloping surface. Use a white pencil on the highlights and darken them down near the radii with the Steel followed by Cool Grey markers. Highlight in the usual way with gouache.

Metals

In their polished state metals behave much like chrome, except that, depending on the metal, they do not reflect colour to the same degree: reflections in polished brass, for instance, are highly contrasted yellows and browns. Stainless steel and aluminium reflect more colour but with a greyish blue tinge while copper reflects pinks and reddy-browns.

If you need to render any of these (particularly if it is an all-metal product such as silver and brass ware) then collect a few examples and pictures of similar items before starting. I recommend using an Ingres paper of the same basic colour as the metal, and

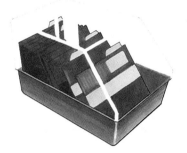

1

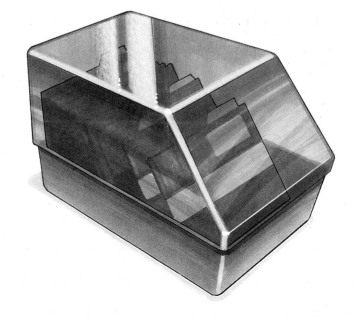

2

preferably one that approximates to a mid-tone, and then using a marker to put in the key reflections. After this use light- and dark-coloured crayons for highlights and shadows.

Heavy Textures

As with the example of the car interior illustrated on p. 129, heavy textures can usually be simulated by slipping a textured surface beneath the paper and overworking with a coloured pencil, much as you would with a brass rubbing. Often a realistic three-dimensional effect can be obtained by using a light pencil first, and then moving the texture slightly and overlaying a dark pencil.

Coarse and fine sandpapers are useful (especially for fabrics and carpets) and some materials, such as wire mesh and perforated steel, can be simulated using the actual material behind the paper. If you are airbrushing, these can also be used for spraying through to create a textured effect and, of course, save on mask cutting.

10 Descriptive Drawing

Apart from full colour presentation visuals, there are many other types of drawing that a designer may be required to produce. The most obvious, and demanding, are the orthographic technical drawings that cover everything from GA (General Arrangement) and parts drawings to fully measured perspectives. There are many books on technical drawing and it is also one subject which is well covered by design education, and so is not dealt with here.

Of the other types, most are usually done to describe a design concept further in terms of its construction or use, or simply to establish scale, or to illustrate the product in use.

Cut-Away Drawing

Doing accurate cut-away drawings can be a difficult and laborious process that demands very careful construction. For the designer, absolute technical accuracy is probably unnecessary; on the other hand, a cut-away view can be very descriptive of how a product works, or is put together. The best compromise between accuracy and clarity is achieved by being very selective about *which* parts are cut away, and by how much. Concentrate on revealing those parts which really need description, and avoid showing those which do nothing to add to the viewer's understanding and also take a long time to draw. It is usually possible to overlap revealed areas in such a way that the one nearest to the viewer obscures all but the *key* parts of the one beneath, and, in this way, saves drawing time.

It is absolutely vital with a cut-away to get the basic perspective correct. With a normal visual you cannot, of course, see anything of the other side of the product, and the furthest unseen (but inferred) corner can be grossly misplaced before the view starts to look slightly odd. With a cut-away, however, if you do not start correctly, the perspective errors become more obvious as you lay in interior detail. The drawing should be built up in the normal way, blocking in the major shapes first, and roughly locating key points on the inside with respect to the outside (most easily done if you are working from a fairly resolved GA). Some designers prefer to begin on the inside and work out to the exterior shell, but this can involve you in a lot of unnecessary work. I prefer to firm up on the outside so that the product looks complete and correct, and then begin to sketch in the interior detail, cutting away and revealing as I go. The tightening up of the line-work is left until I have finally decided on that which is to be cut

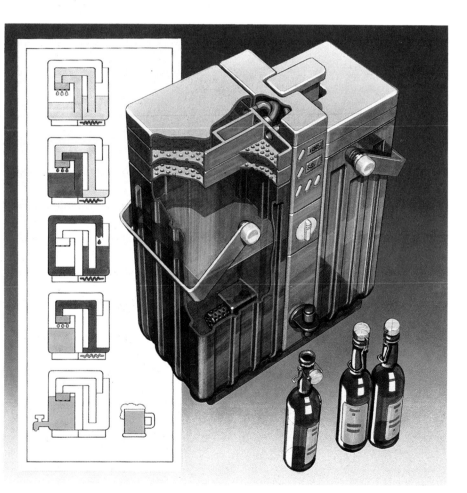

Freehand Line Drawing

Freehand drawing has more character and is quicker to do than a drawing done with rulers and guides, which can look very sterile. It is also useful to give a conceptual feel to the design.

Below: Agricultural machine
For General Electric Plastics

Like the beer-making machine, this drawing was done for reproduction in a calendar and shows an idea for an agricultural power-pack. The final drawing was done freehand with technical pens. At the layout stage, however, the drawing was fully developed using straight-edges and guides, and these were only discarded for the final inking. Note the graphic effect of having elements breaking out of their surrounding box, such as the man's foot, and the flexible drive cable.

A clean sheet of tracing paper was laid over the underlay and the image traced through freehand using two sizes of technical pen. A 0.7mm was used around the perimeter of each major part to establish that it was in front of the one behind, and a 0.3mm was used for all the other lines. To add depth to the drawing, an intermediate line thickness was introduced for falling-away edges. Finally, some good quality dye-line prints were run off, and the drawing coloured with crayons (although any medium can be used).

away and revealed, and that which is not.

Once the line-work is complete, the drawing can be finished using any technique. For straight descriptive work, particularly if the drawing is for eventual reproduction, a superb effect can be obtained using commercially available dry-transfer tones. These are available in a wide variety of percentage tints, dot and linear patterns, and graded tones. If it is important to separate graphically and identify one revealed part from another, with no requirement for realism, then completely flat colours (also available in stick-down sheets) can be used either with or without tones. If you have the time, then you can render the complete cut-away machine, as here, as if it were a real cut-away product of the type found in a science museum, an impression strengthened by the bright red cut-line.

Left: Beer-making machine
For General Electric Plastics

I wanted to indicate how the machine might work and so chose to do this cut-away view which shows the filter trays clearly. The main parts of the drawing were done with a series of overlaid brown markers; note how, to get the transparent effect of the smoked polycarbonate, you can see the base and handle in silhouette through the right-hand container. The top surfaces were done with brown pastel, and a white crayon was used to model the other parts. The bright red cut-line was initially put in with a red pencil, and then overlaid with red gouache. The simple flow diagram illustrating the brewing cycle provided a nice background behind the main product and both these drawings were cut out and mounted to the airbrushed background. The beer bottles were done in a much looser and bolder style, with a classic window highlight on the upward-facing surface.

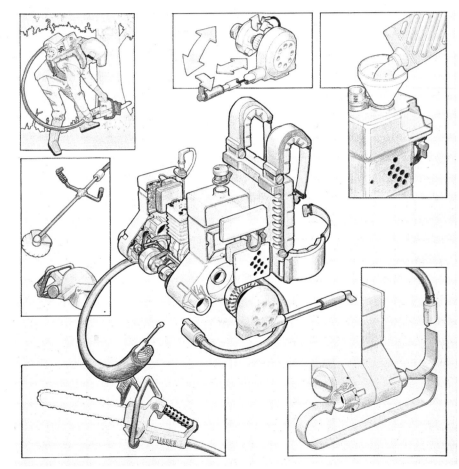

Exploded View

Few designers get involved in really accurate exploded views because they can take a very long time to construct. However, there are plenty of occasions when a schematic exploded view helps to explain the disposition of parts and mouldings, and the general layout of the product. If you have already constructed a three-quarter view, then this can be used as the basis for an exploded view, either by expanding the perspective or by moving the underlay in the appropriate direction and tracing off each part in turn.

Above right: Generator
For Yamaha Motors N.V.

In this example, the generator internals were stock parts and so these were photographed, traced off and enlarged. The various mouldings were then sketched in on this basic underlay.

In deciding which part should hover where, you should also consider the function of the drawing; if, as here, it is merely to indicate where plastics might substitute for steel, then the panels can be overlaid one on another to give depth and save on construction time. Once the view was resolved, a clean sheet of tracing was overlaid and the underlay traced off with a 0.3mm technical pen; the perimeter of each part was then outlined with a thick Overhead Projection pen to make it stand out from its neighbour. Good quality white dye-line prints were taken, and the appropriate panels coloured with blue markers and modelled and highlighted in the usual way.

Right: Induction cooker and ultrasonic dishwasher
For General Electric Plastics
by Seymour/Powell

This axonometric view was drawn onto film using technical pens and self-adhesive dot tones and then a 5× 4in black-and-white photograph taken and printed onto resin-coated paper. This was then airbrushed with flat colour with no modelling of the parts (except the control panel detail). Care is needed when using inks on photographic paper because excessive build-up of ink will lift off easily with subsequent overmasking.

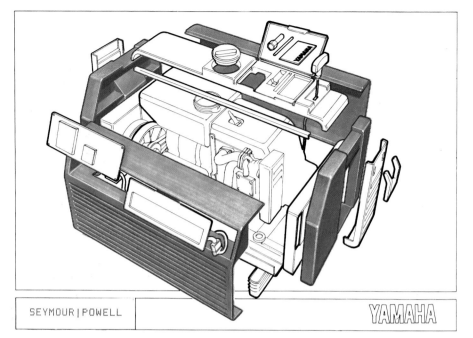

SEYMOUR | POWELL   YAMAHA

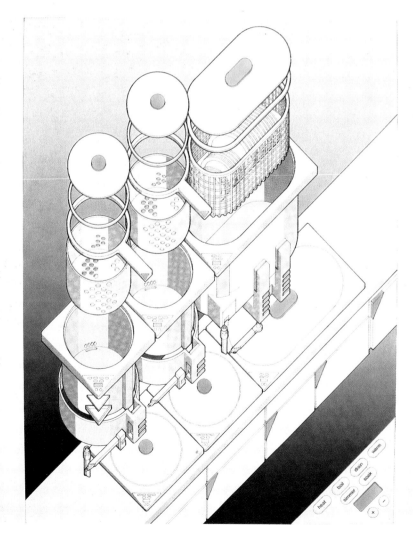

Rendered General Arrangement Drawing

The designer's work is not all concept and ideas – indeed, much of his time is taken up in the resolution of the idea into a realizable product. Inevitably, once the model-making stages are underway the designer depends less and less on his presentation skills because he no longer needs to create the illusion of three dimensions. Often, however, there is a phase after the initial concepts, where the chosen theme is developed through to a General Arrangement drawing. During this phase, the final layout is decided, dimensions are fixed, ergonomics resolved, production methods finalized and so on; this is the real meat of the design process where the designer juggles with all the conflicting factors to perfect the final solution. The decisions made during this phase need to be discussed with the client before a finished model can be commissioned, and often it is sufficient to present only the finished GA drawing. Sometimes, however, as a result of development, the concept has changed slightly and the client may therefore need to see the implications of these changes in a more visual way, particularly if he is unaccustomed to reading technical drawings.

In these situations you could, of course, do a new visual of the design, but the traditional visual does not reflect the detailed nature of the phase that has been completed. The solution is to render up the finished GA drawing (provided there are sufficient complete views) in the chosen colours and complete with graphics. You can use any technique for this if you work on prints taken from the original tracing, alternatively you can work directly on the tracing with the advantage that subsequent prints come out beautifully modelled not in colour, but in monotone.

Designers who are very skilled at this type of presentation use it much more interactively than simply colouring up a finished GA and can therefore combine the design/search phase and final presentation within a single drawing.

Tractor control console
For the Massey-Ferguson Manufacturing Co. by Peter Ralph

The drawing was done by overlaying heavy-grade tracing film over the full-size concept layout for the cab and controls. The theme was evolved progressively in pencil, ink and airbrush, and repeated corrections and adjustments were made with reference to engineering 'hard' points, problems of appearance, ergonomics and manufacturing feasibility. The hard surface of the film allowed for repeated erasing without breaking up the surface which might happen with art board or paper.

The technique is a flexible, continuous process, in which the designer can present ideas to himself, selecting and rejecting, until a rounded, overall concept appears.

Because it conforms to the established engineering design communication process, the drawing can be understood by, and disseminated throughout, the client organization – both to management for corporate assessment, and to engineering for detail development.

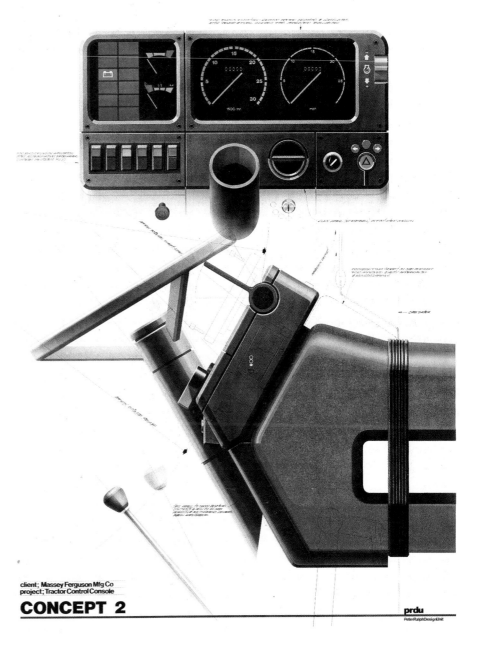

client; Massey Ferguson Mfg Co
project; Tractor Control Console

CONCEPT 2

prdu
Peter Ralph Design Unit

Machining centre
For the Newall Engineering Co.
by Peter Ralph

In this example, the proposed design was traced down onto Frisk CS10 Artboard with the aid of a drafting machine and then airbrushed in the normal way with gouache and inks. The process was greatly speeded up, and accuracy maintained, by using the drafting machine when cutting masks. This does, however, produce a high turnover of plastic drafting scales!

Primary chaincase
For Norton Motors Ltd.

One aspect of a design job for Norton Motors was to restyle this primary chaincase so that it looked less like something off a 1960s BSA and more appropriate to the high-technology Wankel engine lurking behind it. The job did not allow us to modify the case's mating surface, as this would have incurred heavy re-tooling costs. To ensure that the proposed design would fit, the engineering drawing for the old chaincase was used as an underlay for the new design ideas. In this way, all the bosses and centre-lines were correctly positioned and we ended up with a full-size view of the proposed design.

The drawing is a working drawing, and one of several alternatives generated. It was drawn out with a fineline black marker first, and then coloured over entirely with a Cool Grey 8 marker; the darker areas were put in with a Black. The Norton roundel was heavily pastelled from top to bottom with white, and then the name cut back in with the grey marker to produce a slight shadow. The edges were then picked out with a black and a white pencil. The red background square was done with a Cadmium Red marker edged with a Venetian Red, which was also used to suggest a shadow beneath the case.

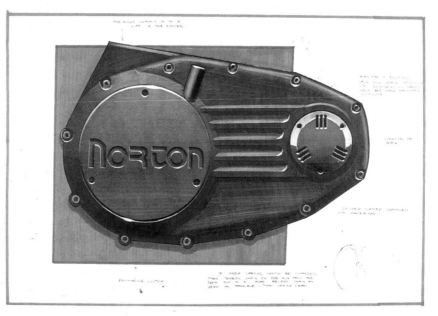

Rendering as Model

As a fast and effective means of getting a concept into three dimensions the card model is hard to beat, particularly if the product has many flat panels. Modern foamcore cards are rigid, light and easy to use, and so allow the designer to build up an informative image of the product very quickly. A sketch model is not, however, a finished model and sometimes, if it looks finished, it can mislead the client into believing that the concept is a great deal more advanced than it really is. On other occasions, for example when producing two or three alternative ideas at an early stage, adding realism to the model can do much to remove the starkness of the white card which is usually a feature of sketch, or space, models. Indeed, when designing generators for Yamaha, I have used the technique of laminating renderings to the model to establish detail and provide the illusion of slots, fixings, control panels etc. Obviously, this is only possible if you are working on scale drawings such as the GAs on the previous pages.

Truck cab
For Leyland
by Peter Ralph

In this example the model was rendered in the flat onto Frisk CS10 Artboard, and assembled much like a cut-out from the back of a cereal packet. The folded card 'capsule' was then planted onto a wooden chassis raft and wheels. Note how the grilles, wipers, grab handles and even shadows were all rendered to enhance the three-dimensional impression.

Concept Drawing — Software

Nearly everything in this book is hardware of some sort or another because this is what, by and large, the product designer is preoccupied with. The graphic designer and advertising creative will also find much of this useful when doing scamps and roughs. Much of their work, however, is spent drawing the rest of the world — for example, people, hands, faces, fabrics, trees, landscapes, etc. They nearly always work from reference (unless endowed with a photographic memory) and often complete the line-work in pen before colouring up. As with all presentation drawings, it is the quality of this base drawing which, more than anything else, determines how good the finished visual will look.

For the product designer, this type of visual is often used alongside a product drawing, as a research board for marketing purposes. The product is clearly shown and the image, or lifestyle, suggested in the background.

1

2

3

4

WHEN YOU DON'T NEED TO BE
TOO MUCH OF A LADY....

5

Research board
by Richard Seymour

1. The line drawing is laid out with a black Ball Pentel and a black Pantone fineliner around the perimeter. Note how the hair breaks out of the background and how the left hand, which will be left predominantly white, bleeds out into the background. The face is to be lit from head-on, as if illuminated with a flash bulb; this is very difficult to do well because it washes out all the contrast and reduces shadow areas which help define shape.

2. Burnt Sienna marker is applied across the hair as a basic foundation which can be either darkened later by further modelling with the marker, or lightened with a white pencil. The swirls of the hair are followed with the marker to emphasize their shape. Barely Beige marker is used as a foundation for those areas of the face and hands thrown into shadow: under the hairline, down the sides of the nose, under the nose, under the cheekbones and under the lip. The bottle is treated with a

Cadmium Yellow marker leaving highlights along the top radius and in the thick glass at the bottom. The enamelled gold top is put in with a Light Suntan marker leaving a broad white highlight down the centre.

3. The background is put in with an Africano marker; when the drawing is finished, half of this will be her fur coat, which will be picked out by backlighting the edge. The same marker is used to darken the undersides of the curls and provide depth. The facial shadows are further modelled with a Light Suntan marker. Burnt Sienna marker is used around the bottle top highlight to add contrast to the reflection, and this is blended in using the original Light Suntan to avoid leaving a hard line. The highlight in the glass is treated in the same way, as is the base where you are looking through very thick glass at the table beneath.

4. The eyes are put in with a Pale Blue marker leaving a highlight dead centre. Because the eyes are reflecting light from directly in front, the resulting highlight is

completely obscuring the dark pupil behind. Cadmium Red is used on the lips, overlaid with a Venetian Red for the shadow areas (on the upper lip, and not quite at the bottom of the lower lip – this dark area will flank the lip highlight applied at the end.) A Pthalo Blue marker is used to make a tiny shadow on the eye (thrown by the lid) and on the lower rim (thrown by the eye itself). A white crayon is then used on all the front-facing curls, for the lip highlight, and for the fur coat. The lettering in the glass of the bottle is filled in with a brown marker.

5. White gouache is used for the highlights around the edge of the coat, in the eyes, the lower lip and sparingly in the hair. It is a good idea when doing this to half-close your eyes in order to pick out a few key points and avoid 'overdosing' the drawing with highlights. Finally, a Pale Blue marker is used around the perimeter of the drawing as if it were casting a shadow.

1

2

3

Advertising rough (1)
by Richard Seymour

1.   As with the previous example, the drawing is completely laid in with a black Ball Pentel and a black Pantone fineliner around the pack itself.

2.   A Pale Blue marker is laid across both sky and mountains, a Light Green for the hills, and a Barely Beige for the rock outcrop and background.

3.   The mules have a base colour of Light Suntan with Burnt Sienna overlaid to produce the shadow areas; the two are then blended before they have time to dry completely, using the Light Suntan. An Africano marker is used for the deep shadow areas such as those cast by the head, the traces, and on the underside of the body, etc. Yellow Ochre is used for the traces with Dark Suntan overlaid where they get darker. Warm Greys are used for the wagon and a Nile Green to model the background crudely. The pack itself is put in with Cadmium Red, and the men roughly coloured with the same markers as for the mules; a Pale and a Pthalo Blue are used for the man's shirt.

4.   A white crayon is used on the mountains, and generally for highlighting; the side of the pack is shadowed with Venetian Red and Cool Grey 1 markers. Finally a sheet of tracing paper is lightly sprayed with Spray Mount, and tacked down across the whole drawing; the foreground areas (including the lettering) are then cut away so that the colours are at full strength in these areas, and they therefore stand out from the background.

Below right: Advertising rough (2)
by Richard Seymour

This is another example of the technique used in the previous examples but with different lighting. Note, in particular, the use of blended markers and white highlights for the skin tones, and the use of a Pale Blue marker for the shadow areas of the white overalls.

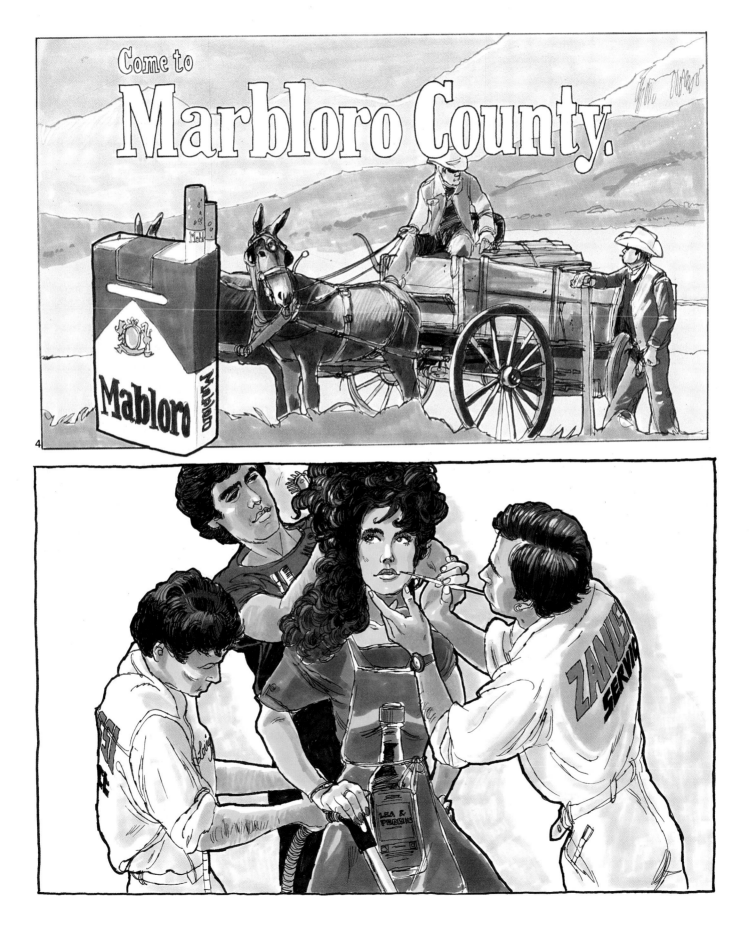

11 Backgrounds and Mounting

If you have put a lot of effort into producing a superb rendering you must ensure that you maintain the standard right through to the final client presentation. Too often designers produce an adequate rendering and then fail to put those finishing touches that can transform the adequate into the exceptional; remember that the quality of your presentation reflects on the quality of the work that precedes it. A shoddily presented drawing can reflect badly on the design work you have done and, equally, a superb presentation can do much to lift what may otherwise be less than perfect work. No matter which, the quality of your overall presentation reflects on you – if you are a professional who is good at your job and have high creative standards then you will settle for nothing less than the best. We have all seen how, when the work of different designers working on the same job is put up for comparison, those concepts which are beautifully presented invariably start with an edge when the decisions are made. Rival designers construe this as an unfair advantage and it may well be that their ideas are better but, if they do not ensure that their work is put forward in the best light, then they have only themselves to blame.

Format and Consistency

If you have several concepts based around a single theme, or several alternatives for a single product, it is a good idea to present them consistently. Where possible, choose one board size (probably one of the A sizes) and stay with it even though this may mean cutting large drawings down and doubling up with the smaller drawings. Also, if possible, choose either landscape or portrait and stick to it. If you are using some sort of background, then ensure that it is the same throughout, or use a simple coloured line or detail to provide a visually linking feature that ties the drawings together. Many designers and design groups evolve their own trademarks which become a 'signature' used in all presentations; this is not necessarily the company logo but can be as simple as a graphic flourish on the end of a line, or a particular type of background. If the drawings are referenced in a report, then they should be numbered for identification; this is a good idea anyway, because it helps during discussions when people refer to this or that concept. Always make sure that your name or the company name appears somewhere on each board – you never know who might see the drawings eventually, perhaps long after the person who commissioned them has left the company, and they might want to know the name of the talented designer who did the work and how to get hold of him. The best way of doing this is to either have a rubber stamp made up, or print a sticker which can be used on the back of each board.

Composition and Backgrounds

Once you have decided on the format, and assuming that you have not done so already, you will need to consider whether to remount the drawing to some sort of background. It is certainly adequate, and sometimes even desirable, to present the drawing simply on a blank sheet of white paper. On the other hand, you can usually do a great deal more to improve the overall image, if only to 'settle' the drawing on the paper.

Organizing the image(s) on the surface to produce a harmonious and graphically balanced board is the most important aspect of the final presentation, after the drawing itself that is. Even if you are simply repositioning the drawing on a fresh sheet of white paper, you should consider carefully just *where* on the paper it should go. It is as an aid to this composition that backgrounds really come into their own. Everyone knows how a mediocre painting can be transformed by a decent mount and frame, so consider first why this happens. Careful study of the microscopic movements which the eye makes when studying a picture (or even a document such as this page) has shown that the human eye scans the perimeter of the image first, before moving onto the detail. This is why the general layout of a letter is so important and can make such an immediate impression on the reader. By tidying up the perimeter line and retaining the eye within it, the picture frame works in the same way. Similarly with your rendering, if you simply plonk it onto a sheet of white paper, you leave the eye with the meandering perimeter line of the image and nothing to stop it wandering off. Equally, the meandering silhouette does little to relate to the sharply defined rectilinear edge of the paper or board, and so sits uneasily on the paper.

An appropriate background can do much to make the product look more dynamic and exciting, or perhaps simply emphasize a particular line or direction (as with the motorbike on p. 76). It can assist the viewer's understanding of the product by placing it in some appropriate context that illustrates where, or how, it is used. A background can also do much to make the image look more three-dimensional, either by throwing it forward, or making it appear to sit on a surface. Equally it can be used to tie a series of dissimilar images together on a board.

The possibilities can be broken down into two main types: composition and technique, although, of course, they can both be used together.

Compositional elements
The picture frame. This is a simple line drawn inside the paper edge which, working just like a mount or frame, retains the eye and settles the drawing.

A shape. A rectilinear shape, either simple line or block colour, behind the image not only throws it forward, but can also visually tie the irregular shape of the drawing to the edge of the paper or board.

'Seating' the image. Often it is better to have the product apparently sitting on a surface rather than floating around in space. This can be achieved by actually drawing the surface or by simply laying in a shadow beneath it. With the former, tiles (sometimes reflective) are very popular as they are easy to draw, and car designers in particular like to show their vehicles sitting on reflective surfaces so that a mirror image is seen in the ground. If the viewpoint is very low so that the horizon is on the surface itself, the underlay can be simply flicked over to produce the required image (see example on p. 130), otherwise some more accurate construction is called for.

3 - W H E E L   R E C R E A T I O N A L   V E H I C L E

Concept drawing for a sports three-wheeler
For General Electric
by Richard Seymour

This is an example both of a realist background (here applied to a quick concept drawing) and of an image seated on a surface. The view was intended to be a pastiche of a well-known rendering by Syd Mead, but since the concept was rejected, the sketch was never progressed through to a finished visual. Note how the tiles are reflected in the wheels to make them look chrome, and also how pale blue crayon has been used to create the white bodywork.

Realism. You can, of course, set the product in a complete environment, drawing everything and anything around it. This is effective, but time consuming; often it is sufficient to suggest elements of the environment that are more economical to produce.

Lifestyle. Particularly when doing concept boards for research, there is a need to show the product in use, or simply to suggest the kind of people and activities associated with it. This kind of lifestyle background requires considerable drawing skill and if you are not very confident it may be better to opt for something simple. The obvious shortcut is to find some photographic reference which can be enlarged and traced off or, better still, take your own photograph. By doing this you obviously have a lot more control over what is being shown in the background than by simply selecting the images from magazines or other pictorial sources.

Background techniques
There are many examples of different backgrounds throughout the book which you can refer to; the following is a summary of techniques, and includes some that are not illustrated.

Graded. Graded papers are available commercially, if somewhat expensively, and add an effective professional finish to drawings. If your budget won't stretch, or, more likely, the colour required is not available, then they are easily airbrushed; using an airbrush also allows you to be more adventurous and produce graded tints which change in colour. With simple masking you can add to the effect – for example, use dry-transfer lettering (perhaps the product name or model number) in the background before you apply the graded tone. After airbrushing, remove the lettering, which has been acting as a mask, to reveal the underlying colour in the shape of the letters.

Streaked. Much loved by auto designers but declining in popularity, this effect can be achieved in several ways: by using the pastel and solvent method (as for wood in the preceding chapter), by using giant markers and solvent-based ink, or by simply using the felt from inside the marker. With all of these methods, you will need to either mask or work on a separate sheet of paper and then trim out. Curiously they always look a little messy (as opposed to dynamic) if they are not worked to a crisp edge, if, in other words, the turn-around area at the end of each pass is left in the drawing.

Right: Rubbing down a magazine or newspaper picture

1. Choose your picture and soak liberally with a solvent like Flo-master. If the paper is relatively impervious, do this on the facing surface of the picture. Lay the picture face down on the paper.

2. Rub the back of the picture hard with the back of a spoon – lift a corner now and again to check that the image is transferring. Peel off the original and allow to dry.

Below right: Lawnmower
For General Electric Plastics
by Seymour/Powell

The main view was rendered with Magic Markers and coloured crayons and the exploded view drawn with technical pens onto CS10 paper. The photograph was covered with two layers of polyester drafting film so that the overlaid mower stood out. Before final sticking down, a 'drop-shadow' was airbrushed on the back of the film to make the main view 'sit'.

Rubbed-down picture. If you can find a suitable background picture in a magazine or newspaper, a ghosted impression of the original can usually be transferred easily so that the viewer has a hazy impression of the subject. This can be done by dissolving the printing ink with a suitable solvent as shown in the illustrations above. Before you do this, however, take another example from the same magazine or newspaper and run tests to find the solvent which best dissolves the printing ink – paraffin works well for most newspapers, but can leave an oily finish.

Tracing paper overlay. As with the example illustrated right, a photograph can be easily 'knocked back' by overlaying it with tracing paper or film. Sometimes a rendering can be lost against a strong background. By spraying a light coat of glue on to a sheet of tracing paper, laying it across the whole board and then trimming around the product image to remove the paper locally, the main drawing can be re-emphasized.

Spatter finish. This is achieved by loading some ink into an old toothbrush and wiping a knife across the bristles so that spots of ink are thrown onto the paper. Mask well!

Batik. Some interesting batik-like effects can be obtained by using wax to resist the application of coloured inks. Try wiping a plain white candle in broad strokes across the surface before using a giant marker.

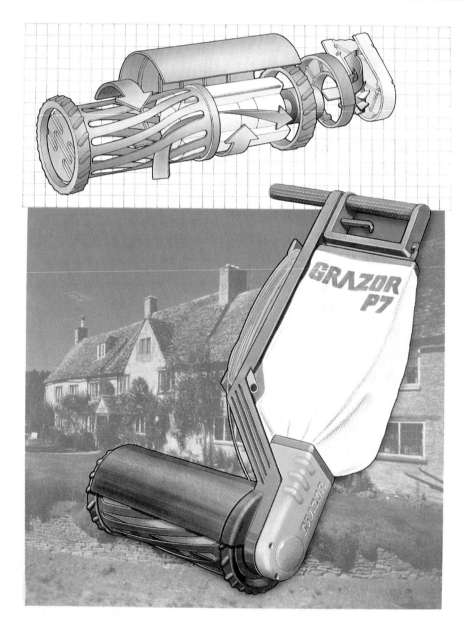

Producing a 'filmed' colour finish
1. Submerge the paper in a shallow tray of water and quickly squirt your chosen colour onto the surface of the water.

2. Reach through the water and draw the paper up through the film so that the colour is transferred onto the surface of the paper.

3. Immediately lay the paper onto a wooden or laminated drawing board and stick all four edges with gum-strip (not masking tape), so that, as it dries, it will stretch and therefore remove unwanted cockles.

'Filmed' colour. This is the finish often seen on the inside covers of old books – a sort of marbled finish. It can be achieved by floating the inks on the surface of a bowl of water and pulling a piece of paper through them to pick up the colours (see illustrations above).

Mounting

Mounting a finished rendering to a sheet of card has a subtle effect on our perception of it – somehow it assumes more authority than its unmounted rivals. Be careful, because this means that a client can view mounted sketches as 'more final' than unmounted ones. Generally speaking, however, mounting a rendering produces only positive benefits. Firstly, because the added whiteness behind the colours can be seen through semi-transparent layout papers, it makes the colours more punchy. Secondly, it smoothes out all the wrinkles, creases, and general unevenness that the paper has suffered through constant handling. And thirdly, the final board is easy to handle and prop up at the client presentation.

On the surface it seems the simplest thing in the world to stick a sheet of paper to a board, but the process is full of potential disasters for the inexperienced. It's a hard way to learn, but once you have ruined a complicated drawing (or worse – one of your boss's drawings!) then you will not make the same mistake again. These though are a few tips to help you avoid some of the pitfalls.

Boards

You must remember that, as boards will inevitably bend during handling, the paper stuck to them (if it is on the outside of the curve) must stretch, because the outside of the curve is longer than the inside. The glue must also be able to accommodate this movement if, when the board is relaxed, you are to avoid the drawing bubbling off. The only way to achieve this is to dry-mount the drawing so that there is an unbreakable, consistent and yet moveable bond between the two. The alternative, of course, is to choose a very rigid board that will not bend and, in terms of traditional artboards, this means using an eight- or ten-sheet construction which is both heavy and expensive.

Recently on the market, however, are the foamcore boards, which are just very expensive. These are exceptionally light and immensely rigid and, better yet, can be cut with a single pass of the scalpel. They are available in a variety of thicknesses up to about 10mm; in some parts of Europe these extra-thick boards can be found in builder's merchants where they are sold as wall insulation and are, consequently, much cheaper than in graphics stores where they tend to be premium-priced. Foamcore is also very useful to the designer as a modelling material for mocking up sketch models.

When cutting board, use a sharp scalpel or knife, and a steel-edged rule, and, if you are not absolutely confident, always put the rule on the side of the work and the scalpel on the side of the waste, so that if you slip you won't ruin the drawing. If you are cutting through thick artboard, get the cut established first with a couple of gentle passes of the knife, before applying any pressure. Finally, always mount the paper down first, before trimming the board, so that you have no problems aligning the two and can trim both board and paper together.

Glues

Most popular by far are the spray-glues which are aerosol-based. They are popular because they are so easy and quick to use; this is particularly important if you are mounting a cut-out drawing which has an irregular perimeter – one quick spray and the whole lot is evenly coated with no risk of the glue getting on the wrong side. (Before the advent of these glues most designers used rubber-based cements which were applied with a spatula; as you worked your way around the back of the paper, there was considerable risk of your inadvertently sliding the front of the paper with the drawing over some glue which, because of the need to have glue right to the very edge, had been overwiped onto the baseboard). It is a tribute to their effectiveness that so many designers continue to use spray glues despite the considerable risk to their health. The tiny atomized particles of glue hang in the air and can be easily inhaled into the lungs where, of course they stick. If you continually spray against the same backdrop you will notice how the overspray builds up on it quite quickly and, to a far lesser extent, this is what could be happening to your lungs. Eventually, someone will perfect an electrostatic glue applicator that ensures that all the atomized particles are attracted to the nearest earth (the paper) and therefore not left hanging in the air, but until they do, be sure to take sensible precautions. Try and set up some sort of spray area or cabinet which is well ventilated (preferably with an extractor) and always use a mask.

There are two popular types of spray glues, one fairly light-tack that allows repetitive repositioning (Spray Mount), and one heavy-tack (Photo Mount) which gives a firmer bond but little scope for repositioning. I always use the former for mounting small cut-out drawings, and the latter for mounting the completed sheet to the mounting board.

There are, of course, alternatives to spray glues, such as the rubber cements (Cow Gum) and dry-mounting. Neither are as quick and both have their problems. Some rubber cements, because they are solvent-based, can actually dissolve any marker inks there may be on the reverse side of the paper and start spreading the colours around! With dry-mounting, you should test all the materials which you use to check that they are undisturbed by the high heat needed to melt the glue. Under these conditions, marker colours can lose their sharp definition and look slightly fuzzy.

Both rubber cements and spray glues can be loosened with lighter-fluid, but check first that this does not adversely affect any of the other materials. (This, incidentally, is one of the disadvantages of using foamcore board because the lighter-fluid is absorbed by the porous foam and stains the surface before it can evaporate). Finally, watch out for sharp changes in environmental temperature and humidity – excessive damp or very hot conditions can make the paper expand and bubble off. Indeed, in very hot conditions (in a studio with a glass roof and inadequate ventilation), I used to keep A3 renderings awaiting presentation in the fridge!

Protecting the Drawing

Having mounted the rendering to a piece of board you should consider how best to protect the surface in a way that enhances the overall quality of the presentation. At its most basic, a simple fly sheet of coloured paper taped or glued to the back of the board and folded across the front will perform adequately. If the sheet is like a thin card (such as the high-gloss Astrolux or

Chromolux) then always score the fold lightly so that the cover does not bow. Clients do have an annoying habit of jabbing their fingers into the finished drawing to make some point or other and leaving a terrible mark or, worse, smudging the pastel. Of course, it doesn't really matter by then, but it is a little irksome when you have just poured your all into the creation of a beautifully finished drawing. The solution is to overlay a sheet of protective film which protects the drawing from such attack, and allows offending marks to be removed.

If you have the time, the most professional way to do this is to have the drawings laminated. In this process, the drawing is sealed between two sheets of clear flexible plastic, with all the air squeezed out under pressure. If you decide on this approach, then do *not* mount the drawings to card as there is a limit to the thickness which the machine can accommodate; simply remount them to another sheet of white paper to bolster the colours. It is sometimes possible to have the backing sheet of black, rather than clear, plastic which can look cleaner, but should only be used if there is no risk of dulling the colours. You can laminate in a variety of thicknesses of plastic with the thinnest being a lot cheaper than the thickest; the process gives the drawing considerable rigidity, so that it will prop up against a wall, and looks extremely slick. Laminating does not seem to affect the drawing in any way, although it is always a good idea to run a test first to make sure. It is also absolutely irretrievable once done, so be sure that you photograph the drawing first and that it really is finished. Because it gives a completely sealed surface, you can write notes and comments on the surface and wipe them away – this can be quite a useful technique if you wish to point out key features of the design but do not wish to permanently disfigure the drawing. One great advantage of laminating is that it seems to offer good resistance to fading brought about by exposure to ultra-violet light. Nearly all types of colouring media fade eventually although some colours are more fugitive than others. Markers, in particular, are not really intended to last for centuries and fade quickly, so if you want to keep drawings always cover them and store in a drawer away from light. Laminated drawings, on the other hand, seem to last much better; indeed some examples which have been framed, and hung in a light room for three years, show no discernible deterioration of the colour (as yet).

An alternative to laminating is to use a sheet of acetate lightly glued to the surface of the drawing (see adjoining illustrations for

how to do this). The main advantage of this method is its accessibility; there is no need to visit the local laminators, or pre-book their time, with the attendant risk that their machines may be out of action. There is also the (remote) risk that they will contrive to mangle your precious artwork in the rollers and, of course, they may not be available at 2 o'clock on the morning of the presentation when you need the work done! The acetate protects the drawing, and offers some UV protection; it is also removable if you need to photograph the drawing or make some alterations.

A third way of sealing the surface of the drawing with a film of plastic is heat-sealing. This is usually done in a dry-mounting press with both pressure and heat and is therefore only suitable for drawings which are undisturbed by the heat – be sure to test first. With marker drawings the process can blur the edges of the colour quite badly so that the drawing begins to look a little fuzzy; also, if you have used white paint or crayon on top of black marker then the process might make these areas mauve rather than white!

One final alternative is the traditional fixative which can be sprayed onto the surface of the drawing; these are now quite sophisticated products which can be built up layer by layer like varnish to give a good finish and adequate protection. They are easy to use, but somewhat unpredictable: as with heat-sealing, they can turn white paint on black marker to a nice mauve colour. On the other hand, they can often give a sharp lift to the colours, so, as with the other methods, test first. Some companies are now offering UV-retardant fixatives (e.g. Marabu) which have a significant effect on the life of the colours.

Presentation

The professional is not only concerned with *what* he is presenting, but how he presents it. Every designer knows the importance of this but there are those who do it with style and panache, and those who fumble nervously through it and do nothing to give the client confidence in the work being presented. If you are presenting good quality work that you are proud of, then the task will be easier than if you are trying to paper over the cracks of a less than perfect job. It is a shame, however, if the work is good but you simply spread it out on the table for the client to leaf through haphazardly. It is difficult to offer any constructive advice to those who find it hard to stand up in front of others and talk through their work because it is mainly an issue of self-confidence and confidence in the work

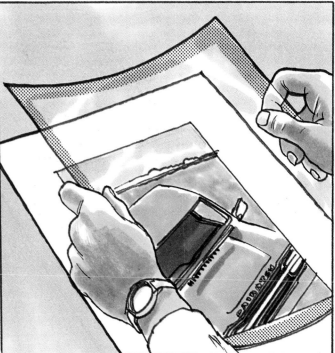

Using acetate sheet to protect a visual
1. Choose a fairly heavyweight acetate, preferably from a pad rather than a roll so that it lies flat and does not curl. Cut a sheet of layout paper about 10mm smaller than the acetate all round and lay it in the middle so that only a 10mm edge of acetate is visible. Spray this lightly with Spray Mount.

2. Remove the paper mask and carefully lay the acetate over the rendering. Use a soft brush to apply pressure all round and ensure a good bond. Unlike laminating, the protective sheet can easily be removed at a later date if necessary.

you have done. It may be useful, though, to look at the two main methods of presenting work, the gradually unfolding story and the overview, so that the inexperienced have at least some starting point on which to build.

The unfolding story
With this method the designer controls what the clients see, and the order in which they see it. The designer can therefore apply an order, chronological or otherwise, that shows the work in the best light, and, just as important, the clients have nothing to distract them – they must listen to what you say rather than allow their thoughts to dart ahead to the next drawing. This method allows you to take the clients through your thinking and illustrate *why* you made the decisions you did; they can see how the idea developed, how you modified it, what you were striving for, what you rejected and so on, and are thus better able to understand the final concept and are probably a great deal more appreciative of the time it took to do the work (and therefore less likely to flinch at the size of your bill!). I usually assemble the best of the

preceding work into a sketchbook with a spring binder, and mount only the final concepts to board so that they can be pinned, or propped, for comparative purposes.

Some designers photograph all the work and present it as a slide-show rather than as drawings – this can have much impact but also makes it difficult for the clients to check back to something they saw earlier. For this reason, a slide show should always be backed up by the real work so that discussion is possible afterwards.

The overview
This method really demands that you have access in advance to the room where the final presentation will take place. You can therefore pin all the work up in the most advantageous way so that when the clients enter they see everything at once. This offers the maximum impact 'up-front' and the clients can see immediately how much work you have done and its overall quality. However, even though you may take them through the work, the element of surprise is lost and they may become restless because

they are eager to progress to some later stage they saw when they first entered the room. It is also more difficult to present preliminary sketch work because the sketches tend to assume the same weight as the rest of the presentation when given the importance of a place on the wall, and, anyway, there is usually insufficient room for all the back-up work.

Both methods have their advantages and the final method is usually decided by the type of work you have produced, and the place where it is to be presented. Whichever you choose, it is a good idea, if you have the facilities, to reduce all the drawings down to A4 and add them to your final report in a neat bound booklet, ensuring that you have enough copies for everyone who will be at the meeting. This allows every person at the presentation to carry away a complete record of the work you have done, which they can use during project discussions in situations where the finished boards might prove clumsy and unwieldy. Be sure that you only give this out *after* your presentation (to prevent them jumping ahead).

12 Conclusion

Most of the techniques described in this book are those which I personally have used and developed since I completed my further education. Nearly everything was initially learnt from someone else and then developed into something which I could begin to call my own. I had always been interested in drawing but became more so while working alongside car designers at the Royal College of Art. Their techniques were eminently suitable for the cars and trucks which they were designing but less so for the consumer products that I needed to draw.

Immediately upon leaving college I was thrown in at the deep end and required to render a huge variety of things from woodgrained electric fires to brass reproduction lighting – it was a baptism of fire that forced me to take the bare bones of what I had learned at the RCA and make it work. Each new job became a challenge as I struggled to draw clear mouldings, pressed brass, mahogany veneer, or whatever the project demanded. Gradually a style emerged that I could call my own, but this was not just a formula applied like a colour-by-numbers picture. If there is a style that other people recognize as mine, then it is evolving constantly because every drawing represents a new problem: it may just be rattled off, but more likely, as I learn from colleagues, and as new materials become available, I will experiment before taking refuge in a tried and trusted technique.

If you are already a designer but have no particular aptitude for, or interest in, rendering and therefore have difficulty surmounting those occasional presentation problems, this book will still be a useful source of reference. If you are a beginner, don't be afraid to copy from others. In some colleges, this is (surprisingly) rather frowned upon – it won't help you to design, or draw a good perspective, but it will help you understand the materials you are using and how to get the best from them. Look through the magazines that regularly carry examples of renderings (such as the excellent *Car Styling*) and try to reproduce a drawing. As you try to figure out how the designer achieved this or that effect and then emulate it yourself, your mind will be darting ahead, working out how to do it next time, and most important, how to do it better.

One of the factors which will encourage you to change and develop your techniques is the development of new graphic materials – just as the marker has almost completely replaced gouache and other brush-based media, so new materials constantly offer new opportunities for the designer. It is important, therefore, to keep in touch with the latest developments.

If you enjoy drawing, then be sure to build on and advance beyond what is here. Rendering is not a religion, there is no dogmatic way of doing this or that drawing; take what you can from these pages and then, through experimentation and learning from others, move on to develop your own techniques.

Index

Figures in *italics* refer to illustrations

A

acetate sheets 20, 62
　to protect a drawing 154, *155*
acrylic 23, 24
　rendering 139, *139*
adhesives 23, 24, *25*, 152, 153-4, 155
advertising roughs *148, 149*
air sources for airbrush 20, *21*
airbrush rendering 13, 20, 25, 86-8, *89-92, 93-5, 96-7, 98, 99, 100, 101*, 139
　freehanding 88
　lines 86, *87*
　masking 86, 88, *88*
　toning 88
airbrushes 9, 20, *21*
architecture 6
auto design 110, 118, 127
automotive rendering 20, 110, *110-12, 113-15, 116-17, 118-26*, 127, *127-9, 130, 131, 132, 133, 134, 135, 136-7*

B

backgrounds to finished work 150-53
　batik 152
　'filmed' colour 153, *153*
　graded paper technique 151
　lifestyle 151
　picture frame 150
　rubbed-down picture 152, *152*
　shape 150
　'seating' the image 150-51
　splatter finish 152
　tracing paper overlay 152
backlighting 59
batik 152
bleed-through 13, 62
bleeding 13, 16, 60
board brush 24, *25*
boards 13, 88, 150
bridge 24, *25*
brushes 9, 20, *21*
bulldog clip 14, 16

C

Callum, Ian (rendering by) *127-9*
candle 152
car exteriors 10, 109, 110, *110-12, 113-15, 116-17, 118-26*, 127, *130-31, 132, 135, 136-7, 145, 151*
car interiors 127, *127-9, 133*
carbon tetrachloride 24
Catignani, Tony (rendering by) *116-17*
chalk 25
'chings' *see* highlights
chrome finishes, reflections in 49, *49, 50-51, 52-3*, 53, *54*, 55, *55-6*
Chromolux 124
circle in perspective 32, *32, 33*, 34, *34, 35*
　to divide up 36, *36*
　see also ellipses
cleansers *see* solvents
colour
　finish with pastel 18, 19, *18-19*
　flat finish 13, 16, 60, *61*, 88, 102
　pull 16, *16*
　knocking back 24, 62
　see also reflections, rendering
coloured paper rendering 102, *103-5, 106, 107, 108, 109*
coloured papers 13, 102, 138
coloured pastels 13, 18, 19, 24, 25, 60, 102, 110, 138, 151
　to obtain broad fields of colour *19*
　to obtain a smooth finish *18-19*
coloured pencils 9, 13, 17-18, *17*, 27, 47, 60, 102, 110, *116-17*, 138, 139
　using *18*
colouring up 48-59
compasses 23, *25*, 26, 88
compound forms in perspective 39
computers, use of 9, 28
concept drawing 146, *146-7, 148-9, 151*
cotton wool 16, *16*, 18, 19, *19*, 25
Crystalline paper 62, 119
cubes
　in perspective 28, *28, 29*, 30, *30-31*
　reflections in 49-50, *49, 50-51*, 57-8, *57-8, 59*
　rendering with airbrush 88
curves 23-4, *24*
cut-away drawing *44, 80*, 140-41, *141*
cutting mat *25*
cylinder
　in perspective 34, *34*
　reflections 52-3, *52-3, 54-5*
　rendering 60, *61*, 88

D

darklight *see* lowlights
descriptive drawing 140-49
　concept drawing 146-9
　cut-away *44, 80*, 140-41, *140*
　exploded view *82*, 142, *142, 152*
　freehand line 141, *141*
　model rendering 145, *145*
　rendered GA drawing 143, *143*
desert cliché *52*, 53, *53*, 54, *54*, 55, 56, *56*
double horizon 119
drafting work 13, 23
drawing materials 12-25
drawing skills 6, 9-11, 26, 27, 39, 140-44
drawings, presentation of 155
drawings, protecting 16, 154, *155*
dry mounting 154
dry transfers 75, 141, 151
dyeline prints 48

E

electrostatic spraypen 20
elevations 39, 65, 73, 96, 100, 110, 138
ellipse guides/templates 24, *24*, 28, 37, 141
　to construct a view *34*
　to construct circular shapes *35*
ellipses 32, *32, 33*, 34, *34*, 35, *35*
　changing value of *34*
　properties of 32, *32*
English, Jim (rendering by) *107, 108*
episcope 28
explanatory view *85, 98*
exploded views *82*, 142, *142, 152*

F

Farey, Cyril (rendering by) *7*
felt wadding 14, *16*
'filmed' colour background 153, *153*
finishes, reflectivity
　chrome 49, *49*, 50, *50-51*, 52-3, *53, 54*, 55, *55-6*
　gloss 49, *51*, 53, *53, 54*, 55, 56
　matt 49-50, *51*, 53, *53*, 55, *55, 56*
finishes, rendering
　metals 102, 114, 139
　textured material 139
　transparent materials 49, 63, 102, 138-9, *139*
　wood 138, *138*
finished rendering 62, 102
fixative 16, 68, 154
foamcore boards 13, 145, 153, 154
freehand line drawing 141, *141*
freehand underlay 42, *42-3*
freehanding (with airbrush) 88

G

General Arrangement (GA) drawing 26, 40, *40-41*
General Arrangement, rendered 143, *143*, 144
giant marker 14, 25, 60, 151, 152
 to make 16, *16*
gloss finishes, reflections in 49, *51*, 53, *53, 54*, 55, *56*
glues *see* adhesives
gouache 20, *21*, 86
graded paper technique 151
Grant enlarger 28
graphite 25
grid overlay 28
gum strip 153

H

Harvey, Charles (rendering by) *135*
health hazards 20, 23, 153-4
heat-sealing 154
highlights 16, 18, 19, *19*, 20, 48, 50, *51*, 56, 57, 59, *59*, 102, *116*, 139
horizon line *26*, 27
Hotblack, Marcus (rendering by) *133*
Hutchinson, Peter (rendering by) *132*

I

Iijima, Yukio (rendering by) *130-31*
industrial design 6, 9-11
Industrial Revolution 6
Ingres paper 13, 102, 139
inks 13, 20, *21*, 24, 86, 151, 152, 153
Ital design *109*

K

'knocking back' colour 24, 62

L

Lai, Pinky (renderings by) *131, 133*
laminating to protect drawings 16, 154
layout papers 13, 16, 25, 62, 155
lettering 20, 42
lifestyle background 146, 150
lighter-fluid 24, 88, 154
lines 13, 20, 24
 with coloured pencil 18, *18*
 with airbrush 86, *87*
 with marker *15*, 62
lint 16, *16*, 18, 25, 62
liquid mask 25
Loewy, Raymond (renderings by) *8, 9, 10*
lowlights 114
Lutyens, Sir Edward 7

M

marker ink 16, *25*
marker paper 16, 60
marker rendering 13, 60-63, *63-4, 65, 66, 67-9, 70-72, 73-5, 76-7, 78, 79, 80, 81, 82, 83, 84, 85, 110-12, 113-15*
 lines *15*, 62
 flat colour 13, 16, 60, *61*, 102
 masking 60, *61*
 streaking 60, *61*, 138, 151
 toning and blending 62
markers 9, 13-17, *14, 15*, 27, 138, 154
masking 13, 23, 60, 86, 88, *88*, 91, 96, 139
masking film 23, 25, 60, 86, 88, *88*
masking tape 13, 16, 23, *25*, 60, *61*, 62
matt finishes, reflectivity 49-50, *51*, 53, *53*, 55, *55, 56*
Mead, Syd 151
Melanex 85
Melville, Ken (renderings by) *110-12*
 and Gunvant Mistry *118-26*
metals, rendering 139
methylated spirits 20
Midland Bank, London *7*
Mistry, Gunvant and Ken Melville (rendering by) *118-26*
model making 11, *145*
models 11, 20, 28, 118, *145*, 153
mounting of finished work 13, 23, 150, 153-4
Mylar 49

O

Ogle Designs *109*
 and Charles Harvey *135*
 and Jim English *107, 108*
 and Marcus Hotblack *133*

P

PMT machine 28
PVC sheets 23-4, 62
 to make a straight-edge 23, *23*
paints 9, 20-21, 60, 102
 gouache 20, *21*, 86
pantograph 28
papers 13, 16, 25, 60, 62, 102, 110, 119, 155
pastel and solvent method 18, *19*, 138, 151
pastels *see* coloured pastels
pencils 9, *12*, 13, 24, 25, 102, 139
perspective 6, 26, 27, 39, 110, 140
perspective drawing 26-7, *40-41, 42-3, 44, 45, 46, 47*
 circles 32, *32, 33*, 34, *34, 35*, 36, *36*
 compound forms 39
 cubes 28, *28, 29*, 30, *30-31*
 cylinders 34, *34*
 radiussed edges 38, *38-9*
 spheres 36, *36-7*
perspective grid 28, 39, 40-41
photocopier 28
photographs 28, 152
picture frame background 150
plan-view 96
polyester film 23, 49, 119
presentation drawing 6-9, 11, 13, 20, 146
presentation of work 150, 154-5
products illustrated:
 action vehicle *134*
 advanced transportation concept (magnetic levitation vehicles) *133*
 agricultural machine *141*
 airball game *106*
 bathroom unit *45, 99*
 beer-making machine *44, 140*
 bottles *65*
 camera *42-3*
 car interior *127-9*
 central-heating pump *79*
 circular knitting machine *108*
 compressor *107*
 cutlery *96-7*
 DIY knives *66*
 disc storage box *139*
 ECO 2000 concept car *136-7*
 electric drill *70-72*
 electric radiant heater *10*
 electric toaster *10*
 electronic calculator *78*
 electronic calculator with printer *78*
 electronic movie camera *89-92*
 fan heaters *8, 9*
 four-door saloon *130-31*
 four-wheel drive ATV *132*
 generator *73-5, 142*
 glue-gun *46*
 glue-gun and soldering iron *80*
 grill *107*
 haircurler set *84*
 hairdryer *40-41*

incinerator *109*
induction cooker and ultrasonic dishwasher *142*
infra-red optical reading head *100*
lawnmower *152*
light fitting *98*
livery for BR Freightliner container lorry *11*
machining centre *144*
mahogany table *138*
mechanical digger *63-4*
modular process control system *101*
motorcycles *76-7, 81*
motorcycle controls *84*
motorized bicycle wheel *83*
pedal car *83*
pencil sharpener *103-5*
personal computer TV *78*
plastic saddle *82*
plug *8*
portable computer *78*
portable tool-storage unit *85*
primary chaincase *144*
roadrunner truck *135*
safety helmets *100*
Sony Walkman *47*
sports car *10, 110-12*
sports car (the Maya) *109*
sports three-wheeler *151*
telephone *93-5*
three-door family car *116-17*
tractor control console *143*
.truck cab *145*
truck interior *133*
two-door saloon *131*
two-seater city commuter *118-26*
two-seater convertible *113-15*
vacuum cleaner *67-9*
vacuum flask *85*
'X-ray squad' wrist unit *47*

R
radiussed edges
  in perspective 38, *38-9*
  reflections 56, *56-7*
Ralph, Peter (renderings by) *143, 144, 145*
realist background 150
reflections 74, 75, 79, 103, 130, 139
  cube 49-50, *49, 50-51*
  cylinder 52-3, *52-3, 54-5*
  radiussed edges 56, *56-7*
  sphere 55, *55-6*
  to construct in a flat surface 57-8, *57-8*
  to make a surface look reflective 58, *59*
rendering skills 6, 9, 11, 26, 48-9, 60-63,
  86-8, 102, 138-9
research boards 146, *146-7*
rubbed-down picture 152, *152*
rubber cements 153, 154
rubbers 13, 18, 24, *25*
rulers 18, 20, 23, 24, 62, 87, *87*

S
sandpaper 139
scale, to indicate 26, *27*, 73
scalpels 16, 18, 23, 24, *25*, 88
scriber 23, 24
seating the image 150
Seymour, Richard (renderings by) *146-7,
  148-9, 152, 153, 155*
Seymour/Powell (renderings by) *100, 142,
  152*
shadows 48, 59, *59*, 139
sharpeners 13, *25*
shut-lines 16, 20, 26
'signatures' 150
single light source 48-9, 59, *59*
size of drawing 27
sketch models 28, *145*, 153
sketch rendering 62-3, 102
sketching 13, 20, 26-7, 39
solvent inks 24, *25*
solvents 16, 17, 18, 19, 20, 23, 24, *25*, 62,
  152
spheres
  in perspective 36, *36-7*
  reflections in 55, *55-6*
  rendered by airbrush 88
splatter finish 152
spraymarkers 20
steel rule 23
straight-edges 23, 62
  to make *23*
streaked background 151
streaking, marker 60, *61*, 138, 151
Stevens, Peter (renderings by) *107, 113-15*
Sturges House, California *8*
Sugar paper 102
sweeps 23-4, *24*

T
talcum powder 18, 19, 24, 62
tape-drawing 20, 23, 118, *118-26*
tapes 13, 16, 23, *25*, 60, 61, 62
technical drawing 6, 23, 26, 73
technical instruments 22, *23*
technical pens 22, *23*, 141
textured materials, rendering 129, 139
Thompson, Julian (rendering by) *135*
tissue paper 19, 24, 25, 62, 88
toning and blending 17, 18, 62, 86, 88, 102
tracing-down paper 25, 102
tracing film 13
tracing off 47, 67
tracing paper 13, 23, 25, 62, 110, 119
tracing paper overlay 152
transparent materials, rendering 49, 102,
  138-9, *139*
turned finish 76, *76*
Tustin, Don (renderings by) 11, 100, 101

U
UV-retardant fixatives 16, 154
underlays 13, 25, 26, 39, *40-41, 42-3, 44,
  45, 46, 47*, 48, 102, 150

V
Vellum 62, 110, 119
  rendering on *110-12*
views
  choosing 27, *27*

W
washing-up liquid 116
wet-and-dry paper 23
'wet front' 16, 60, *61*
window cliché 53, 58
wire mesh 139
wood, rendering 138, *138*
Wright, Frank Lloyd (rendering by) *8*

Y
Yu, Scott (rendering by) *136-7*